Absorption and Emission
by Atmospheric Gases

Absorption and Emission by Atmospheric Gases

The Physical Processes

EARL J. McCARTNEY

John Wiley & Sons

New York / Chichester / Brisbane / Toronto / Singapore

Library of Congress Cataloging in Publication Data:

McCartney, Earl J., 1908–
    Absorption and emission by atmospheric gases.

    (Wiley series in pure and applied optics, ISSN 0277-
2493)
    Includes bibliographical references and indexes.
    1. Meteorological optics. 2. Gases—Optical
properties. 3. Spectrum analysis. 4. Absorption of
light. 5. Emission spectroscopy. I. Title. II. Series.

QC975.2.M4 1983   551.5'6   82-17443
ISBN 0-471-04817-8

Printed in the United States of America

10 9 8 7 6 5 4 3 2 1

To my parents

Jesse and Sarah McCartney

Preface

We live in a radiative environment whose source is the solar energy intercepted by the earth. Captured and transformed by the earth's surface and atmosphere, the primary stream of solar energy initiates and supports a myriad of secondary streams. Carrying energy at all wavelengths and moving in all directions through our terrestrial domain, these radiative streams create a temperature environment that makes possible all life on earth. Basic in this capture and redistribution of radiant energy are the processes of atmospheric absorption and emission.

A proper treatment of this subject requires that a large amount of information be selected, organized, and interpreted. Knowledge of atmospheric physical and radiative properties has grown apace in recent decades. The use of airborne platforms for sampling and measurement has provided a wealth of directly gathered data on atmospheric constituents and properties. Radiometric probing has yielded new information on trace gases and the vertical distribution of the atmosphere. Orbiting satellites carry radiometric and photographic probing to the ultimate, and meteorology has become a truly global science. The steady development of infrared sensors and systems has made this technology an indispensable partner in military weaponry and aerospace projects. In all of these activities atmospheric absorption and emission play significant roles and often are basic factors.

I have written this book for those who wish to understand better the processes of absorption and emission and their manifold effects. Persons having such interests or needs are the workers in meteorology, atmospheric physics, aerospace surveillance, and air-pollution control. The potential audience may consist of scientists, development engineers, technicians, and inquiring students of the environment. I have had these prospective readers in mind because I am one of the group. As an electro-optical engineer for many years, I had the early good fortune to share in the development of sea-going infrared systems. My involvement with infrared and with the larger subject matter of this book is a long-term affair.

The objectives of this book are to bring together the essential theories and factual data of absorption–emission, to interpret and present this material in a systematic manner, and to introduce the reader to some of the voluminous literature on the subject. Practically the same statement with regard to scattering was made in the preface to *Optics of the Atmosphere:*

Scattering by Molecules and Particles, which should be considered a companion volume to this one. Here also I strive to mediate between specialized knowledge in the fields of atmospheric physics and spectroscopy and the needs of persons who, although versed in technology, are not specialists in these fields. An intermediate level of presentation has been employed as being suitable for this purpose, in view of the expected diversity of reader backgrounds. From the wide fields encompassed by the title of this book, and the nearly bewildering amount of data, I have tried to select those portions that will help to educate the readers, aid them in defining practical problems, and encourage them to go further.

The atmospheric context of absorption and emission are emphasized by frequently reminding the reader that these are processes occurring in the atmosphere. A reasonable and desirable next step would be to consider *radiative transfer* and *earth's heat balance,* which are the atmospheric results of these processes. Closely allied thereto is *inversion of measurements,* which allows the structure of the atmosphere to be inferred from remote measurements of its radiance. From this it is but a step to *laser probing of the atmosphere,* a most promising development. Thus the progression grows. Regrettably, none of these matters could be covered in this volume because of considerations of book size and desired coherence of topics. Therefore, only absorption and emission as processes in themselves are examined.

In the preparation of this book I am indebted to many persons. The idea for such a work originated during my early infrared experiences at the Sperry Gyroscope Division of the Sperry Rand Corporation. Under the direction of Fred Braddon, Victor Vacquier, and Richard Scheib, Jr., I was encouraged to dig deeper and seek the mechanism behind the phenomena. May every engineer have such opportunities. The volume itself was planned and portions written while I was a research engineer in the Electro-Optics Group at Sperry. Much assistance was received from Robert Jagoe, Leonhard Holmboe, Thomas Hutchison, and Jimmie Aushman.

My thanks go to Ernest Wittke of the Kearfott Division, Singer Corporation, and to George Harvey of the U.S. Naval Research Laboratory for their interest and many helpful comments. I express my appreciation to Beatrice Shube of John Wiley & Sons for expert advice and never-failing encouragement. Above all, I am grateful to my wife Marie for creating an environment in which the writing of this volume has been a pleasant and rewarding task.

EARL J. MCCARTNEY

Rockville Centre, New York
December 1982

Contents

Absorption and Emission
by Atmospheric Gases

1

Introductory Ideas
and Useful Facts

The Atmosphere gives life—in and by the Atmosphere everything has its being—it is the Atmosphere which stores up the solar rays in order to give us the benefit of them in the future and makes us an abode in which we act as if we were the sole tenants of the infinite—the masters of the universe—without the atmosphere our planet would be inert and arid, silent and lifeless.

Thus did the astronomer Flammarion view the air envelope of the earth in 1872. A full century later we cannot state the matter more cogently.

Central to the life-giving functions of the atmosphere are the absorption and emission of radiant energy by the constituent gases. Investigations of gaseous absorption and emission began early in the last century when Fraunhofer and Wollaston analyzed sunlight and found dark lines scattered throughout the bright solar spectrum. Thus came the science of spectroscopy, unexcelled in elegance of concept and precision of measurement. Spectroscopy has grown far from its origin in the visible spectrum and extends across the ultraviolet, visible, infrared, and microwave regions. No techniques of chemical analysis are more sensitive and conclusive than those of spectroscopy. Positive identifications for thousands of molecular species are provided by their characteristic absorptions and emissions. Our best insights into the structures of atoms and molecules are based on spectroscopic data interpreted by the theories of quantum mechanics. Compositions of the sun, stars, planetary atmospheres, and the thin gases of interplanetary space are all revealed by analysis of the light from these faraway sources.

Close to our subject, meteorologists study absorption and emission as the mechanisms that harness radiant energy from the sun and earth to the atmospheric heat engine whose workings are the *weather*. Climatologists deal with these affairs globally and synoptically as the factors that keep the earth's average temperature in balance. Environmentalists, alert to any effects of atmospheric pollution on this balance, can make their missions more realistic by further study of these processes. Environmental problems will never go away so long as human life exists on this planet, but, rightly appraised, these problems can yield to technology and management. Optical system engineers, who design sensing and surveillance equipment for aerospace purposes, must always reckon with atmospheric absorption and emission.

This chapter, which presents some introductory ideas and facts, is intended as background information for the treatments that follow. Several disciplines are spanned but few derivations are given. The statements and equations should be accepted as premises that will be justified in the remaining chapters as needed. Here we provide information for later use and give an advance view of subjects to come. We start with an overview of absorption–emission and proceed to the electromagnetic spectrum with its parameters of wavelength, frequency, and wavenumber. The ultraviolet, visible, and infrared regions of the total spectrum are reviewed briefly. Next, the basics of energy quantization are presented in simple terms. We look at the molecular origins of absorption and emission and at the behavior of spectral lines. The laws of blackbody radiation are then discussed, with emphasis on atmospheric applications.

This book employs the International System of Units, or *SI Units,* that is an extension and standardization of the meter–kilogram–second–ampere (MKSA) system. However, when citing published research findings and data that have been expressed in other systems or the traditional units of spectroscopy, the original units are used herein. In such cases, the equivalent unit often has been stated except when so doing would clutter a discussion or drawing. The SI base and supplementary units are defined in Appendix B, along with a note and conversions for the units of force, energy, and power relevant here. The SI radiometric quantities and the radiometric properties of matter are reviewed in Appendices C and D.

1.1. AN OVERVIEW OF ABSORPTION AND EMISSION

The atmospheric gases display an astonishing diversity of absorption and emission characteristics. These gases usually are classified as *permanent* and *variable,* but the terms are somewhat relative. Over a long time span any of the gases may be variable to some extent. Those in the permanent class, however, have not shown significant yearly variations since synoptic measurement data became available. The permanent gases are listed in Table 1.1 where it is seen from the volume ratios that they account for about 99.97% of the atmospheric bulk. They are uniformly mixed from the earth's surface up to an altitude of about 90 km because of atmospheric motions.

The measurement unit *atm cm* (atmosphere centimeter) in the table defines the total amount of gas along any path, whatever its direction, length, or altitude. Consider that the entire amount of a particular species

TABLE 1.1. ATMOSPHERIC GASES PRESENT IN PERMANENT AMOUNTS

Constituent	Molecular Weight	Volume Ratio (%)	Parts per Million (ppm)	Total Amount in Vertical Column (atm cm)
Nitrogen, N_2	28.0134	78.084	—	6.245×10^5
Oxygen, O_2	31.9988	20.948	—	1.675×10^5
Argon, Ar	39.948	0.934	—	7.471×10^3
Neon, Ne	20.183	—	18.18	14.5
Helium, He	4.003	—	5.2	4.2
Krypton, Kr	83.80	—	1.1	0.9
Xenon, Xe	131.30	—	0.09	0.007
Hydrogen, H_2	2.016	—	0.5	0.4
Methane, CH_4	16.043	—	2.0	1.6
Nitrous oxide,[a] N_2O	44.015	—	0.5	0.4
Carbon monoxide,[a] CO	28.009	—	0.19	0.15

Source: USSA (1976) and Junge (1963).
[a]Concentration varies in polluted air.

in a path of given length is reduced to a layer at standard conditions of temperature and pressure, which are taken as 0°C and 1 atm (1013.25 mbar). Consider further that all other species are excluded from the path. The resulting thickness of the layer is expressed in units of atm cm. It is evident that the cross-sectional area of the path is not a factor, but unit area is implied and will be consistent with later practice. As a reference standard, we note that the total amount of atmosphere in a vertical column extending from sea level to the top of the atmosphere is 7.998×10^5 atm cm. This is called the *reduced height* of the atmosphere. The values of atm cm in Table 1.1 refer to such a column.

The variable gases are listed in Table 1.2. Although their amounts are small, their radiative importance in some cases exceeds that of the permanent gases. The amounts of the variable gases fluctuate between wide limits, as indicated in the table, depending on geographic location, altitude, season, weather, time of day, and proximity to industrial regions. Hence they are not always uniformly mixed with the more plentiful permanent gases, and local information is often needed to specify their amounts along a given path. Carbon dioxide is somewhat of an exception in this regard. Although exhibiting seasonal variations and an annual increase, it is well mixed in the troposphere, which extends from the earth's surface to altitudes near 11 km.

The absorption of radiant energy by several of the atmospheric gases is

TABLE 1.2.   ATMOSPHERIC GASES PRESENT IN VARIABLE AMOUNTS

Constituent	Molecular Weight	Volume Ratio (%)	Parts per Million (ppm)	Total Amount in Vertical Column (atm cm)
Carbon dioxide,[a] CO_2	44.010	0.032	320	2.56×10^2
Ozone, O_3	47.998	—	0–0.07 at ground 1–3 at 20–30 km	0.345
Water vapor, H_2O	18.015	0–2	—	—
Nitric acid, HNO_3	63.012	—	$(0–10) \times 10^{-3}$	—
Ammonia, NH_3	17.031	—	Trace	—
Hydrogen sulfide, H_2S	34.076	—	$(2–20) \times 10^{-3}$	—
Sulfur dioxide, SO_2	64.059	—	$(0–20) \times 10^{-3}$	—
Nitrogen dioxide, NO_2	46.006	—	Trace	—
Nitric oxide, NO	30.006	—	Trace	—

Source: USSA (1976) and Junge (1963).
[a]Amount is increasing at an average annual rate of about 0.20%.

shown generally in Figure 1.1. Each curve is marked by regions of low absorption known as *windows* and by regions of high absorption where the atmosphere is virtually opaque. Each gas has an absorption signature as a function of wavelength. The aggregate absorption due to all the gases is depicted by the composite curve at the bottom. This is the *telluric* absorption spectrum, revealed by observing the sun from the earth's surface, with the sighting path thereby extending completely through the atmosphere. When the sun is at a low elevation angle, the resulting long optical path makes it possible to measure the absorptions of trace gases—often existing as ions and free atoms in the upper air—that would remain otherwise undiscovered. Such observations yield much information about constituents at high altitudes where direct sampling or measurement is not feasible.

Absorption and emission act jointly to moderate the extremes of temperature that would obtain in their absence. During the day these processes shield us from the lethal solar ultraviolet and help the earth's surface to capture a large portion of the incoming solar energy. During the night they enable the earth to retain much of the heat stored during the day, thereby insulating us from the cold of space. The moderating effect, which makes our earth habitable, can be appreciated by considering the *airless* moon. During the long lunar day—lasting about 13 terrestrial days—the sunward

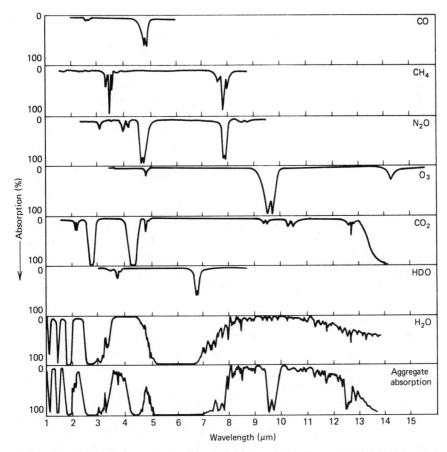

Figure 1.1. Infrared absorptions by atmospheric gases. From *Handbook of Geophysics and Space Environments* by S.L. Valley, Ed. Copyright © 1965 by McGraw-Hill, Inc. Used with permission of McGraw-Hill Book Company.

surface of the moon attains an equilibrium temperature greater than 100°C. The full influx of solar energy, not attenuated by an atmosphere, is balanced at the lunar surface by radiation and reflection, along with slow conduction of heat into the subsurface. At the same time an adjacent region shaded by a terrain feature has a temperature below freezing. During a solar eclipse the surface lies within the earth's shadow for an hour or so, and the temperature falls to about −70°C. Near the end of the long lunar night the temperature may reach −150°C.

Similar extremes of temperature, although not so pronounced, would be experienced on earth were it not for absorption and emission by atmospheric water vapor, carbon dioxide, and ozone. Few subjects are more complex, but a simplified view is appropriate at this stage. Since the last ice age the earth has remained in radiative equilibrium, that is, the average amount of radiant energy sent out to space, over a time span long enough for good averaging, is just equal to the amount received from the sun. The amount sent to space is a function of the effective temperature of the earth–atmosphere system, and this temperature adjusts itself until the outflow is equal to the inflow. When calculated from the radiation laws, the equilibrium temperature turns out to be about $-25°C$. While this is an improvement over the lower range of lunar temperatures cited above it is scarcely conducive to life, except possibly in equatorial latitudes. In pleasant contrast to such a low value, the actual global surface temperature averaged over the seasons is near 15°C. This result is due largely to the trapping of radiant energy by the absorbing gases of the lower atmosphere and is often called the *greenhouse effect,* although the analogy is inexact. Much present concern is expressed that the steady buildup of carbon dioxide in the atmosphere from the burning of fossil fuels may bring about an increase of the average global temperature.

The ozone content of the atmosphere is quite small, but its property of absorbing deadly solar ultraviolet is essential to life on earth. Without this shield, life would have assumed forms far different from those we know, if indeed life could ever have come into being. Although ozone is thinly distributed throughout the atmosphere, it occurs in greatest amount at stratospheric altitudes near 22 km. This is the *ozone layer.* The average amount of ozone—called *total ozone*—in a vertical column from sea level to the top of the atmosphere is only about 0.345 atm cm or $\frac{1}{8}$ in. at standard conditions, yet its absorption is so effective that the lethal portions of solar energy are removed by this tenuous layer high in the atmosphere. We owe very much to very little. Interestingly, atmospheric ozone is continually created and continually destroyed by the solar ultraviolet energy. These contrary processes come into balance at, or maintain, the equilibrium amount of total ozone cited above. Were it not for suspected anthropogenic factors that may tend to destroy ozone, the present equilibrium amount should remain unchanged as long as the supply of solar ultraviolet remains constant.

The fact of atmospheric ozone and its role is a reminder that our total environment is a complex system of materials and processes that came into balance long ago. Within recent years there is concern that the injection of nitrogen oxides into the stratosphere by high flying aircraft, particularly the supersonic transport (SST), poses a threat to the ozone shield. A potential threat also originates at the ground in that chlorofluoromethane (CFM) gases, used as propellants in aerosol sprays, slowly work their way upward to the altitude of the ozone layer. With either type of stratospheric pollution, the resulting photochemical processes in this region of abundant sunlight may decrease the equilibrium amount of total ozone.

Both science and society cannot but agree with Cantor (1978):

It is unnecessary to be an environmentalist to recognize that Mankind's activities have begun to alter what was once regarded, like the oceans, as an infinite, self-cleaning, self-renewing natural resource. The atmosphere has become a dumping ground for the effluents of civilization, with the result that the balance between the sun's radiation, the earth's climate, and the needs of living things, is being altered in significant ways. Whatever the long-term answer to this challenging problem, the short-term requirement is to bring technology to bear on monitoring these changes, and clarifying the chain of reactions that they are capable of initiating. For this purpose, an understanding of the optics of the atmosphere is essential.

1.2. THE ELECTROMAGNETIC SPECTRUM AND ITS PARAMETERS

The electromagnetic spectrum is a representation of either radiant energy or radiant power, displayed as a function of wavelength or a related quantity. Such a display is the necessary framework for all data concerning absorption and emission. An understanding of the parameters that define location and interval within the spectrum will be useful throughout this book. Here attention is called to the essential nature of a spectrum, and three types of spectra are described. Next, the relationships between the three parameters—wavelength, frequency, wavenumber—of spectral location are reviewed. Measurement of spectral interval by means of these parameters is explained. This leads to a discussion of spectral resolving power and Rayleigh's criterion. Typical values of resolving power provided by present spectrometers are noted.

1.2.1. Nature of a Spectrum

The total electromagnetic spectrum is continuous, from gamma rays of vanishingly short wavelength to the 5000-km waves radiated by electric power lines, to still longer waves. This entire expanse covers more than 20 orders of magnitude and transcends all other measurable phenomena. The great differences of wavelength in the various regions indicate a great diversity of sources, but the common factor throughout the spectrum is radiant energy propagating at the speed of light. The spectral regions are distinguished by the types of sources (both natural and man-made), the resulting detectable phenomena, and the types of detectors that can be employed.

Whatever may be the macroscopic form of a light source, the ultimate origins of radiant energy lie in the oscillations of the electric charges that constitute the material of the source. Just as a bullet bears distinguishing marks that enable a ballistics expert to identify the gun from which it was fired, so a specimen of light carries information about its nuclear–atomic–molecular source. When the light is analyzed, its spectrum is found to consist of one of the following:

A series of *sharp lines* that have finite widths.
One or more aggregations of lines called *bands*.
A *spectral continuum* extending over a broad range of wavelengths.

These three types of spectra are illustrated in Figure 1.2. A spectral line means that the energy exists only in a narrow interval centered at a particular wavelength, as in part (*a*). *Line spectra* are characteristic of atomic sources in the form of gases or vapors. The term *spectral line* is an inheritance from that earliest instrument devised for the analysis of light: the prism spectrometer. This instrument displays the differing wavelengths of the entering light as a series of lines, each of which is an image of the entrance slit. *Band spectra,* shown in part (*b*), are the marks of molecular sources in the form of gases or vapors. When examine closely, the band is seen to consist of closely spaced lines, as in the figure. We are mostly concerned with spectra of this type. A *spectral continuum,* as in part (*c*), is characteristic of atomic sources in the form of a liquid or solid, or a gas at very high temperature and pressure. The continuum cannot be resolved because the range of wavelengths is continuous. These three types of spectra appear in both absorption and emission. When the light being analyzed has passed through an absorbing medium, we see *dark* lines, bands, or continua, depending on the selectivity of the absorption. If absorption is absent, we observe the *bright* lines, bands, or continua that characterize the source.

A location in the spectrum is specified optionally by wavelength λ, frequency v, or wavenumber $\bar{v}$. The first two parameters are related by

$$\lambda v = c \qquad (1.1)$$

where c is the speed of light,

$$c = 2.9979 \times 10^8 \text{ m sec}^{-1} \qquad (1.2)$$

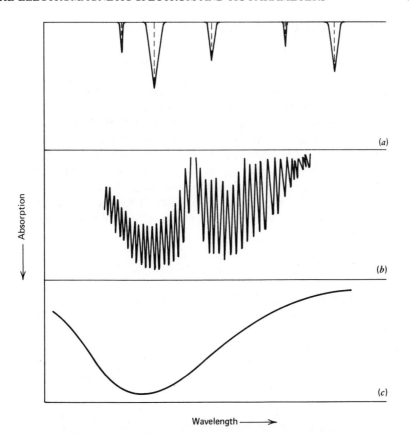

Figure 1.2. (*a*) Line, (*b*) band, and (*c*) continuous spectra.

Optical frequencies are too high to measure directly; for example, the frequency associated with the wavelength 1 μm is 3.00×10^{14} Hz. Because wavelength is measurable with an interferometer it has traditionally been used to define spectral location. Developments of lasers and sensitive detectors, however, enable frequency differences to be measured by heterodyne techniques similar in principle to those long employed in radio.

In spectroscopy a spectral location usually is specified by wavenumber $\tilde{\nu}$, which is the reciprocal of wavelength. At the wavelength 1 μm the wavenumber is 10,000 cm^{-1}, which means that there are this many cycles of wave in 1 cm of length. Wavenumber and frequency are related by

$$\tilde{\nu} = \frac{1}{\lambda} = \frac{\nu}{c} \tag{1.3}$$

When $\bar{\nu}$ is stated in cm^{-1}, the numerical relation is

$$\bar{\nu} = 3.336 \times 10^{-11} \times \nu \ \text{cm}^{-1} \tag{1.4}$$

The unit cm^{-1} is used also for energy, as explained in Appendix B. Occasionally the concept of wavenumber is called *spatial frequency* and sometimes merely *frequency;* we should be alert to such practices.

The characterizing of light by its frequency is useful in much of modern optics and especially in communications. Basically the idea of frequency involves the counting (or naming) of occurrences per unit time. This may be a more primitive operation than the measuring of length, which always requires a comparison in one way or another. The rate of change of frequency with respect to wavelength is found by differentiating Eq. (1.1),

$$\frac{d\nu}{d\lambda} = -\frac{c}{\lambda^2} \tag{1.5}$$

where the minus sign means that the rate decreases as λ increases. The relation (1.5) indicates that the rate of change, per unit wavelength, becomes very large at short wavelengths. For example, at $\lambda = 1.00 \ \mu$m we find from Eqs. (1.2) and (1.5) that $d\nu/d\lambda = 3 \times 10^{20}$. This trend may be inferred from Figure 1.3 where the number of whole frequencies in any decade is equal to nine times the frequency value shown at the lower limit of the decade.

1.2.2. Divisions of the Optical Spectrum

The optical spectrum, illustrated by Figure 1.3, is that portion of the total electromagnetic spectrum to which optical laws and techniques, originally developed for visible light, have been applied. In 1664 Newton produced a visible spectrum by intercepting a beam of sunlight with a prism, thereby demonstrating the color composition of white light. For more than 100 years the idea of a spectrum (from *spectre:* a visible but disembodied spirit) was confined to this little band of colors. Not until Young originated the wave theory in the early 19th century was a clear relation established between the color of light (a perception) and its wavelength (a measurable quantity). About that time other workers extended the techniques used with visible light and discovered radiant energy in the "dark" regions bounding the visible. From these observations came knowledge of the ultraviolet and infrared regions, which, with the visible, constitute the optical spectrum. Rayleigh (1889a) recounts the speculative viewpoints during these formative years, and the belief that the spectrum consisted of

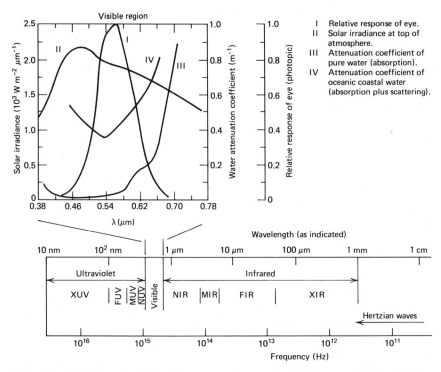

Figure 1.3. Optical portion of the electromagnetic spectrum. Irradiance and attenuation date from List (1966).

three different entities: actinic, luminous, and infrared rays. Here we survey the three divisions of the optical spectrum, calling attention to several features that are relevant to our subject.

The *ultraviolet region* extends from the long-wave limit of X rays at about 4 nm to the violet edge of the visible region near 380 nm (0.38 μm). It is convenient to divide this range of almost two decades into the portions listed in Table 1.3. In 1801 the German physicist Ritter investigated solar ultraviolet by employing silver chloride as a detector, noting that it was darkened quite readily by light in this region. Similar halide materials in photographic emulsions are still used to record ultraviolet, often called *actinic light* because it promotes chemical reactions with many kinds of materials. These reactions are due to the high energies of ultraviolet photons, and the most important ones are those that produce an equilibrium amount of ozone in the stratosphere. As a result of the absorption by ozone at wavelengths less than 0.29 μm and by oxygen at wavelengths less than 0.20 μm, practically no solar energy in these regions reaches the ground.

TABLE 1.3.   DIVISIONS OF THE ULTRAVIOLET SPECTRUM

Designation	Abbreviation	Wavelength Limits (nm)
a. Divisions used in astronomy and atmospheric physics		
Near ultraviolet	NUV	380 to 300
Middle ultraviolet	MUV	300 to 200
Far ultraviolet	FUV	200 to 100
Extreme ultraviolet	XUV	100 to 4
b. Divisions used in photobiology		
Ultraviolet A		380 to 320
Ultraviolet B		320 to 290
Ultraviolet C		290 to 200

Ultraviolet at wavelengths greater than 0.29 μm does reach the ground in varying amounts. Although this ultraviolet does not have the lethal qualities associated with the shorter wavelengths, it is quite active in producing photochemical smog—an especially irritating variety—and several biological reactions. The result of exposing human skin to ultraviolet in the range 0.29–0.31 μm is seen in erythema (sunburn) and tanning. Photobiologists evaluate these effects by means of *erythemal action curves,* such as the standard erythemal response curve in part (*a*) of Figure 1.4. This curve shows the relative efficacy of monochromatic flux, having constant power at each wavelength, in producing erythema. The two associated curves in part (*a*) indicate the anticipated erythemal response to solar ultraviolet at ground level for 0.20 and 0.40 cm of total ozone. In each of the three cases the sensitivity reaches a peak near 0.30 μm and decreases toward shorter and longer wavelengths. Shown in part (*b*) are two curves of solar ultraviolet flux as received at ground level for the indicated amounts of total ozone. The greatest erythemal action occurs in those wavelength regions where the received ultraviolet is changing rapidly with total ozone. For example, when the total ozone goes from 0.32 to 0.24 cm, which is a 25% decrease, the ultraviolet at ground level increases by a factor of nearly five at the critical biological wavelengths.

More serious than the triviality of skin tanning and the temporary discomfort of sunburn is the increasing evidence that links skin cancer to the erythemal action spectrum. Additional results of long exposure to ultraviolet are damage to DNA molecules, which are the genetic material of living cells, and production of mutations in animal cells. It is reasonable to suspect that a reduction in the equilibrium amount of total ozone could bring increases in these biological effects, which may not be immediately

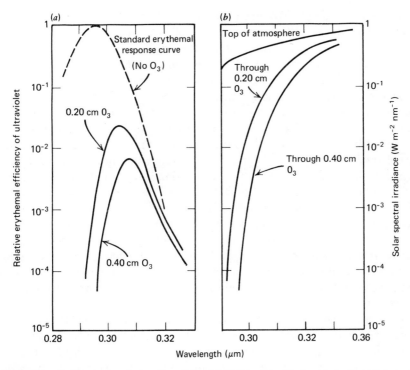

Figure 1.4. Erythermal action and sunlight. (*a*) Erthermal effieciency of ultraviolet at sea level for the indicated total ozone. (*b*) Solar spectral irradiance for the indicated total ozone Date from List (1966) and NAS (1976b).

apparent but tend to be cumulative. Such concerns appear warranted by the curves of Figure 1.4, hence we should be alert to any threat to the ozone layer, from whatever cause. The threats posed by nitrogen oxides and CFM gases, noted in Section 1.1, are less the inevitable consequence of industrial progress than they are the unforeseen result of society's desire for conveniences.

Few books devoted to the ultraviolet have appeared. Koller (1965) surveys the entire region and deals with sources, detectors, applications, and biological effects. The middle ultraviolet is covered by Green (1966) with detailed treatments of all aspects except biological. He provides much information on ultraviolet astronomy, necessarily conducted from rockets and satellites. An introductory account of ultraviolet astronomy is given by Goldberg (1969). A good survey of the SST and CFM threats to ozone will be found in Dotto and Schiff (1978). The views of environmentalists and industrial scientists regarding the CFM threat are presented evenhandedly by Sugden and West (1980). Biological effects are covered for the general audience by Giese (1976), and very thoroughly for the serious student by Parrish et al. (1978) with a monumental bibliography. Much information on the nature of the threats and the potential results are provided by government sponsored studies in FCST (1975) and NAS (1975; 1976a, b).

The *visible region* lies nominally between 0.38 and 0.76 μm, as indicated in Figure 1.3. Whereas other spectral regions are defined by technologies and sometimes arbitrarily, this region is uniquely fixed by the response of the human eye. The limits stated above are only approximate, however, as discussed by Wald (1959). Actually, they are the wavelengths where the visual sensitivity has fallen to about 10^{-3} of its maximum value at 0.555 μm. When high-energy sources are used as stimuli, "... investigations have pursued human vision to about 0.312 μm in the near ultraviolet, and to about 1.050 μm in the near infrared." Nevertheless we follow custom in taking 0.38 and 0.76 μm as the practical limits of the visible spectrum. Considering the great expanse of the total electromagnetic spectrum we marvel that we have been able to gather so much knowledge of the physical world from light in this narrow spectral band. This is a tribute, not only to the information content of the light itself, but also to the processing capability of the human visual system.

Several critical factors come together in the visible region. Referring to Figure 1.3, where the relative response of the eye is shown for comparison, this is the region of maximum solar energy and the maximum transmission of seawater. Except for this region and the radio region at great wavelengths, seawater is practically opaque to electromagnetic energy. The atmosphere also is relatively free from absorption in the visible region so that solar energy, although attenuated somewhat by scattering, reaches the earth's surface in abundance. Here occurs the miracle of photosynthesis, whereby solar energy is used by land plants and algae to convert carbon dioxide and water to carbohydrates—the essence of foodstuffs and, therefore, the basis of life. Here we find phototropism or the bending of plants toward light, the vision of all animals, and the oriented movements of creatures both great and small toward and away from sources of light. These phenomena lie in the domain of photobiology, and it is remarkable that all are centered about the visible region, as indicated in Figure 1.5. Here the quanta of light are matched to the molecular tasks that must be accomplished to create and maintain life. Here the quanta have neither too little energy, as in the infrared, nor too much energy, as in the ultraviolet below 0.30 μm. It seems unlikely that life could originate and proliferate on any planet without light in or near the visible region. Light seems to be a prior requirement for life, as implied by Genesis 1:3.

Few books devoted specifically to the visible spectrum, as such, have appeared. Such a dearth of direct treatment comes about because visible

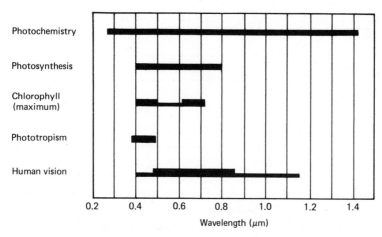

Figure 1.5.   The spectrum of photobiology From Wald, *Life and Light.* © 1959 by Scientific American, Inc. All rights reserved.

light so pervades our lives and activities that the many publications concerned in one way or another with light are not particularly organized around this spectral region, but rather deal with the principles of optics, the behavior of light, and the application of light to the many tasks at hand. Nevertheless, the inquiring reader will find a wealth of ideas and information in SA (1968), where many aspects of light are treated. Those interested in the basics, without a gloss of sophistication, will like Bragg (1940). Many persons will enjoy an outdoor walk with Minnaert (1940) and will learn much about visible light from this master of observation. Somewhat from the viewpoint of spectroscopy, the treatment of visible light by Henderson (1978) is probably unequalled in the literature.

The *infrared region* extends across more than three decades of wavelength, from 0.76 to 1000 μm (1 mm), and for convenience is divided into the four portions listed in Table 1.4. The first three portions contain bands of high and low transmission; these are the results of the selective absorptions shown in Figure 1.1. In most of the extreme infrared, from 15 to 1000 μm, the lower atmosphere is nearly opaque for distances greater than a few meters because of the broadband absorptions by water vapor and carbon dioxide. Owing to these and other absorptions in the range 5–7 μm and at 9.6 μm, the atmosphere acts as a blanket over the earth, preventing the direct escape of thermal energy from the ground and adjacent air. This "greenhouse effect" is infrared phenomenology at its best.

Solar energy at infrared wavelengths was discovered by the English astronomer Herschel (1800a, b), while he was investigating sunlight by moving a thermometer across the visible solar spectrum cast by a prism. He

TABLE 1.4.   DIVISIONS OF THE INFRARED SPECTRUM

Designation	Abbreviation	Wavelength Limits (μm)
Near infrared	NIR	0.76–3
Middle infrared	MIR	3–6
Far infrared	FIR	6–15
Extreme infrared	XIR	15–1000

temporarily laid the thermometer aside, inadvertently in the dark space beyond the red, and then noted that the thermometer still registered significantly. Describing this in his original paper in *Nature,* he apologetically asked permission to use the term *invisible light.* Herschel speculated that this new kind of light differed from visible light only with respect to momentum, a prescient view of photon properties advanced by Einstein a century later. He demonstrated that hot, but not necessarily luminous, objects such as hot stoves emitted this invisible light copiously. From these associations, and the early use of thermal effects for its detection, infrared has been called *thermal radiation* and *heat rays.* We should not suppose, however, that infrared has greater heating capability than visible and ultraviolet light: 1 J represents the same amount of energy anywhere in the spectrum.

Most of the energy emitted by materials at temperatures from a few degrees to several thousand degrees kelvin lies in the infrared. From the energy standpoint, our terrestrial environment is primarily an infrared environment, and this fact the worker in the far infrared quickly learns. He finds that, not only is the outdoor world charged with infrared, but the very laboratory walls and instruments are radiating at the wavelengths he seeks to detect. The situation is analogous to photographing with a camera that has a luminous case.

In recent years infrared has become a major branch of optics. Following World War II, military and aerospace needs have motivated many developments in theory and instrumentation. Several disciplines have strong interests here. The activities range through analysis of starlight and atmospheric constituents, studies of radiative transfer and climatology, remote sensing of earth resources, and development of military surveillance systems. The amount of infrared literature is vast, and numerous tutorial treatments have been published. Among these are Bramson (1968), Button (1978), Hudson (1969), who includes an annotated bibliography of applications, Jamieson et al. (1963), Kingston (1978), Kruse et al. (1963), Smith et al. (1957), Wolfe (1965), and Wolfe and Zissis (1979).

1.2.3. Wave Packets, Spectral Intervals

No specimen of light is characterized by a single wavelength. Even when the light comes from a source called *monochromatic,* it actually is spread over a narrow band of wavelengths. This spectral interval is variously specified in terms of wavelength, frequency, or wavenumber. When the interval is small, Eq. (1.5) can be written in differential form,

$$\frac{\Delta \nu}{\Delta \lambda} = \frac{c}{\lambda_0^2} = \frac{\nu_0}{\lambda_0} \tag{1.6}$$

with the minus sign deleted because a rate is not involved. From Eq. (1.3) the interval in terms of wavenumber is

$$\frac{\Delta \tilde{\nu}}{\Delta \lambda} = \frac{1}{\lambda_0^2} \tag{1.7}$$

Useful relationships from Eq. (1.6) are

$$\frac{\Delta \lambda}{\lambda_0} = \frac{\Delta \nu}{\nu_0} = \frac{\Delta \tilde{\nu}}{\tilde{\nu}_0} \tag{1.8}$$

which link the three parameters.

We note the basic reason why light is not truly monochromatic. Such light would have *constant* wavelength and frequency, unchanging phase structure, fixed linear polarization, and infinite duration. Natural light in the atmosphere is far from meeting these ideals, although laser light comes close. Instead, in observing the light from a macroscopic source we are seeing the resultant of the individual emissions from a multitude of atomic– molecular sources. Even in a low-pressure electrical discharge tube there may be about 10^{16} emitters cm^{-3}. Each emission is random and brief, usually lasting less than 10^{-8} sec, and produces a short wavetrain or packet of waves. A few such overlapping packets are shown in Figure 1.6, where the buildup and decay of the trains signify that the emitters do not start and stop instantaneously. We may regard the emitter as an oscillator that requires a finite time to get going; radiates its stored energy as an electromagnetic wave; and thereby gradually comes to rest. Length of the wave packets is determined by the duration of the emission and the speed of light. In the visual region, say at $\lambda = 0.55$ μm, a packet contains 5.5×10^6 or fewer cycles. A packet is a pulse of very-high-frequency energy.

Due to the brief emitting lifetime, each packet is marked by a spread of

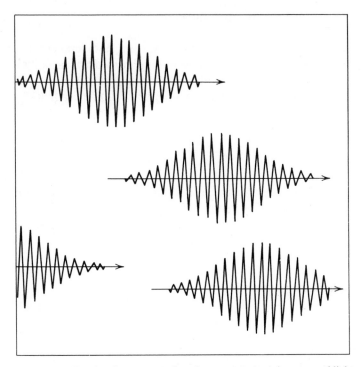

Figure 1.6. Overlapping wave packets from a macroscopic source of light.

wavelengths and frequencies centered about the nominal values of λ_0 and ν_0. This is analogous to the situation in electrical circuits where the frequency spectrum of a pulse becomes broader as the pulse is made narrower. Actually there are two additional reasons for the wavelength spread; all three are discussed in Section 1.4.4. Here we keep in mind that the short lifetime produces a spectral breadth $\Delta\lambda$, defined by

$$\Delta\lambda \approx \frac{\lambda_0}{n} \tag{1.9}$$

where n is the number of cycles in the packet. In terms of frequency this becomes

$$\Delta\nu = \frac{1}{t_0} \tag{1.10}$$

where t_0 is the emitting lifetime.

The spectral breadth of the packets carries over to an equivalent breadth of the resultant wave. From the principle of superposition we know that a number of waves having the same frequency but random phase relations can be added to give a resultant wave of that frequency. The phase of the resultant depends on the amplitudes and phase relations of the components. In this case there are no fixed phase relations among the components because each packet is emitted randomly. Hence the packets are incoherently related. The resultant wave has no regularity of phase structure nor uniformity of period and is the antithesis of a truly monochromatic wave. The "intensity" (standard optical usage but one to be deprecated) of the resultant is equal to the sum of the component intensities.

It is revealing to calculate the spectral breadth of the light from a good quasimonochromatic source. The emission spectrum of the 198 isotope of mercury consists of many sharp lines, some of which are estimated to have a width of only 5×10^{-4} nm. Taking the line at $\lambda = 0.546$ μm, for which $\nu = 5.495 \times 10^{14}$ Hz, substitutions into Eq. (1.8) give a frequency bandwidth $\Delta \nu = 5.032 \times 10^8$ Hz or about 500 megahertz. The resulting ratio of frequency to bandwidth is

$$\frac{5.495 \times 10^{14}}{5.032 \times 10^8} = 1.093 \times 10^6$$

which we may regard as the "quality" or Q-value of the oscillation associated conceptually with the wave. Such a large value argues for a sharply tuned oscillator having a high resonance peak. Application of the Q-value criterion to the narrow lines of gas lasers, which may have bandwidths less than 10 Hz for short periods, yields startling results. Considering the helium–neon laser at 0.6943 μm, for which $\nu = 4.3209 \times 10^{14}$ Hz, we get for the ratio

$$\frac{\nu}{\Delta \nu} = \frac{4.3209 \times 10^{14}}{10} = 4.3209 \times 10^{13}$$

The concept of Q-value was suggested by Lord Kelvin back in 1889 in his studies of oscillatory motion.

1.2.4. Spectral Resolving Power

The concept of spectral interval has practical significance only to the extent that we can measure such narrow intervals and distinguish between adjacent ones. This measuring capability is the foundation of spectroscopy and is expressed by the term *resolving power*. Rayleigh (1879) suggested:

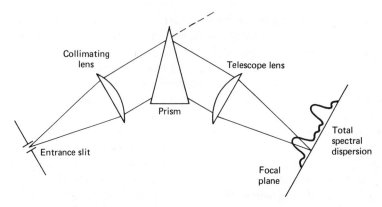

Figure 1.7. Optical system of a prism spectrometer.. Each wavelength produces a diffraction-broadened image of the entrance slit. From *Optics* by F.G. Smith and J.H. Thompson. Copyright 1971 by John Wiley & Sons, Ltd. Reprinted by permission of John Wiley & Sons, Ltd.

"As the power of a telescope is measured by the closeness of the double stars which it can resolve, so the power of a spectroscope ought to be measured by the closeness of the closest double lines in the spectrum which it is competent to resolve." Spectral or chromatic resolving power is defined as the ratio of the wavelength to the minimum wavelength difference that can be distinguished,

$$\text{Spectral resolving power} = \frac{\lambda}{\Delta\lambda} \qquad (1.11)$$

This is just the inverse of Eq. (1.8) but the term $\Delta\lambda$ in Eq. (1.11), instead of having an arbitrary value, represents a measuring capability.

The physical factors that create resolving power are shown in Figure 1.7 where a diffraction grating may replace the prism without affecting the argument. The entering light is collimated by the slit and lens so that plane wavefronts are incident on the prism. By its dispersive property the prism (or grating) changes the wavelength difference $\Delta\lambda$ into an emergence angle difference $\Delta\theta$, which now spatially separates the two adjacent wavelengths. The telescope forms at its focal plane a series of slit images, each corresponding to the entrance slit at a particular wavelength. The image of the slit (a rectangular aperture) is called a *spectral line,* as noted earlier.

Actually the slit image is a Fraunhofer diffraction pattern that consists of a central bright maximum region, flanked by a series of minima and maxima of rapidly decreasing brightness. The separation distance between the centers of adjacent patterns is equal to the product of emergence angle

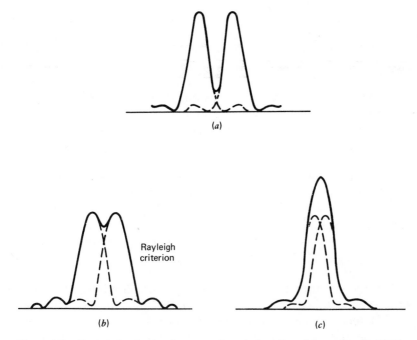

Figure 1.8.   Raylength criterion for spectral resolution. From. Meyer-Arendt (1972).

difference $\Delta\theta$ and the focal length of the telescope. The patterns may be recorded on photographic film at the focal plane or scanned sequentially by a detector moving across the focal plane. Two such patterns are shown in Figure 1.7. Spectral resolving power as defined by Eq. (1.11) depends on the least separation distance at which two adjacent patterns can be distinguished.

A practical measure of resolving power is the criterion proposed by Rayleigh and shown in Figure 1.8. In (a) the two patterns are sufficiently separated to be distinguished easily. An example of the Rayleigh criterion is seen in (b), where the center of one maximum coincides with the first minimum of the other, and the patterns are just resolvable. In (c) the two are so close that they merge into one pattern of increased brightness and cannot be resolved. Employing the criterion in part (b) the quantity $\Delta\lambda$ in (1.11) now can be defined in terms of the diffraction pattern. As stated by Levi (1968): " ... $\Delta\lambda$ is the change in the wavelength λ such that the maximum in the diffraction image of a line, at wavelength $\lambda + \Delta\lambda/2$, coincides with the first zero (minimum) of the image of a line at $\lambda - \Delta\lambda/2$."

Advances in spectroscopy have been marked by steady increases in attainable resolving power. The venerable prism spectrometer, whose concept originated with Newton and which is fittingly displayed on the cover of the *Journal of the Optical Society of America,* gives resolving powers of 10^3–10^4 with visible light. Resolving powers from 10^5 to 10^6 can be obtained with a diffraction grating type through much of the optical spectrum. In both types of instrument the exiting (dispersed) light is spread over a considerable area in the focal plane, as will be understood from Figure 1.7. This limits the sensitivity of the instrument and effectively increases the measurement time. When photographic recording is appropriate, as in the visible and closely adjacent spectral regions, all of the exiting light can be used at the same time. Then light from faint sources can be recorded by employing time exposures, but this increases the measurement time and is unsuitable for rapidly changing events. In the infrared where photographic methods can be used only in a narrow spectral region adjacent to the visible, the spectrum can be recorded by scanning a detector across the focal plane line by line. Since the detector with its own narrow slit aperture sees only one line at a time, most of the light is wasted. In terms of input the sensitivity is low.

The factor of sensitivity or, more properly, the utilization of the entering light is expressed by the term *throughput.* This defines the ability of an optical system to transmit light from entrance aperture to exit aperture. Throughput is equal to the product of aperture area and solid angle of acceptance, and customary units are m^2 sr. The overall throughput is determined by the minimum throughput that exists at any section or aperture in the system. Throughput and resolving power are inversely proportional in any type of spectrometer. This is why the slit width, hence the throughput, must be made small in order to obtain high resolving power in prism and grating type instruments. However, the product of throughput and resolving power does not remain constant in going from one type of instrument to the other; it is somewhat larger for a grating than for a prism type.

In a different approach, interferometers have been applied to spectroscopy with great success. The Fabry-Perot type produces interference fringes as a function of wavelength by multiple reflections from plane-parallel plates which are partly reflecting and partly transmitting. A resolving power of 10^6 and greater is attainable. This instrument is best suited to the detailed analysis of a limited segment of the spectrum that has already been isolated by other means, and it is much used for investigating the hyperfine structure of atomic lines. There is no spatial dispersion of light within the instrument so that it utilizes the entering light efficiently. Readers will recognize that the Fabry-Perot arrangement of parallel, partially reflecting mirrors is used also as the optical resonant cavity of lasers.

The classic Michelson two-beam interferometer, in conjunction with computer processing of detector signals, has revolutionized spectroscopy.

This instrument produces interference fringes as a function of wavelength by differences in path lengths of the two beams. When used as a spectrometer, the path difference is varied steadily while the detector output is recorded as an interferogram. Spectral information then is obtained from a Fourier transform of the interferogram, accomplished with a digital computer. The complete operation is called *Fourier transform spectroscopy* (FTS) and has a resolving power of about 10^6. Because the detector looks at all of the light all of the time, FTS has a throughput that is one or two orders of magnitude greater than can be obtained with dispersing instruments. In employing FTS the user has the option of either getting a good spectrum in less time than with a disperser or of taking a longer time to obtain a better spectrum by means of additional scans.

A well-illustrated introduction to FTS and related matters is given by Connes (1968). In books devoted to larger subjects the essentials of FTS are covered briefly by Smith and Thompson (1971) and by Meyer-Arendt (1972). Spectroscopy by means of grating, Fabry-Perot, and FTS methods is covered by a group of specialists in AO (1969). Similar but monumental coverage of FTS is provided by AO (1978). In addition, Baker et al. (1981) review the development of FTS, including the transform itself, from the initial work by Michelson to present-day rocket-borne instrumentation. Book length treatments of FTS are given by Bell (1972), Ferraro and Basile (1978), Griffiths (1975), Vanasse (1977, 1981), and by Chantry (1971) for long-wave infrared.

1.3. QUANTIZATION OF ENERGY

Nothing demonstrates better the ultimate limitation of a far-reaching concept than the ability of electromagnetic wave theory to explain scattering versus its inability to explain absorption–emission. These processes can be understood only in terms of energy quantization, where the quantum concept is applied to the absorber–emitter and to the electromagnetic energy itself. In this section we review several principles of energy quantization that are needed in later chapters. Of necessity the review is brief, but it may stimulate some readers to recall their earlier studies and encourage others to go further. A wealth of material at all degrees of rigor is available.

1.3.1. Atomic and Molecular Domains

Quantum theory started in 1900 with Planck's law of blackbody radiation. In deriving the law he found it necessary to postulate that the energy of an atomic–molecular oscillator is limited to discrete values that satisfy

$$\mathscr{E} = nh\nu \qquad n = 0, 1, 2, \ldots \qquad (1.12)$$

Here n is the quantum number (always an integer), ν is the frequency of oscillation, and h is Planck's constant having the value

$$h = 6.6263 \times 10^{-34} \text{ J sec} \qquad (1.13)$$

and dimensions ML^2T^{-1}. This quantity is a fundamental constant of nature, and its value must be found by experiment. Very often the quantity $\hbar$, which equals $h/2\pi$ and is read h-cross or h-bar, is used so that a frequently appearing factor of 2π can thus be eliminated. In this book we use only the unbarred h whose value is given above.

Equation (1.12) defines a set of discrete energy values that differ by the value of h when taken in numerical order. The oscillator energy changes abruptly from one permitted value to another, either by absorption or emission of radiant energy or by collisions with other atoms–molecules and consequent exchanges of energy. Such a transition from one energy value (level) to another is the *quantum jump*. This behavior contrasts with the classical view that a continuous range of energy values is available to an oscillator. Quantization of energy is characteristic of atomic and molecular domains. To appreciate this we should note the extremely small value of h in Eq. (1.13), which determines the step-increment or graininess of energy on the microscopic scale. The quantum structure of energy becomes apparent only in dealing with small particles whose energies are correspondingly very small.

The term *oscillator* as used here should be interpreted generically. By this term we mean a particle undergoing periodic motion within a small volume of space. The particle may be an electron bound to an atom or molecule or it may be one of the atomic nuclei that, with attendant electrons, make up a molecule. In citing examples of periodic motion we need to anticipate Section 1.4.1 slightly. In the atomic domain the motion may be that of an electron moving in orbit around the nucleus. In the molecular domain it may be the rotation of the molecule as an entity, or it may be the vibration of a nucleus with respect to the others. The essence of the oscillator concept is that a small particle moves periodically at a very high frequency in a very confined space. Equation (1.12) applies to all situations of this kind. In contrast, the energy of a small particle moving only in free flight (translational motion) is not quantized. Free flight is exemplified by a gas molecule moving unhindered between random collisions with other molecules, and by an electron speeding within the quite empty space of a television picture tube.

Energy quantization of a macroscopic oscillator is not detectable because the value of h is too small relative to the absolute value of the oscillator energy. A simple example from Eisberg and Resnick (1974) is revealing: A pendulum consisting of a 0.01 kg mass is suspended by a string of length $l = 0.1$ m, displaced by angle $\theta = 0.1$ rad, and is then set to swinging. As the amplitude decreases due to friction, can the energy be observed to decrease in discrete steps? From laws of mechanics the oscillation frequency ν_{vib} is given by

$$\nu_{vib} = \frac{1}{2\pi} \left(\frac{g}{l}\right)^{1/2} = 1.58 \; Hz$$

where g is the acceleration of gravity. At the end of the swing the energy $\mathscr{E}$ is all potential,

$$\mathscr{E} = mgl \, (1 - \cos \theta) \approx 5 \times 10^{-5} \; J$$

From Eq. (1.12) the size of each quantized step of the energy decrease is found to be $\Delta \mathscr{E} = 10^{-33}$ J, so that

$$\frac{\Delta \mathscr{E}}{\mathscr{E}} \approx 2 \times 10^{-29} \; J$$

Thus to observe the discreteness of energy for even a small macroscopic system we would require a measurement resolution two parts in 10^{29}, a clearly impossible feat.

1.3.2. Quanta of Light

In employing the idea of quantized oscillators, Planck did not propose that radiant energy itself is emitted in discrete amounts. It remained for Einstein, in explaining the photoelectric effect, to postulate that radiant energy exists and propagates in quantum bits called *photons*. A photon is particlelike or corpuscular in the sense that it is emitted as a short burst of energy during a period of 10^{-8} sec or less. The photon—an entity—then travels at the speed of light, its energy remaining bunched or gathered to itself. When the photon encounters an absorbing atom or molecule, its energy is transferred to the absorber and that photon ceases to exist. A photon must either exist in the full amount of its energy or not exist at all. When a beam of light is attenuated by an absorbing medium, the attenuation is caused, not by reductions in the energies of individual photons, but by the disappearance of whole photons.

Although a photon behaves like a particle, we visualize that associated with it is a short wavetrain having identifiable wavelength and frequency. These attributes we would assign to the light if we were considering its wave aspects, and here also they mark the spectral location of the photon. We regard the wavetrains accompanying the photons as similar to the wave packets of Figure 1.6. The photon is somewhere within this accompanying wave which—if we like—guides the photon in its flight. We cannot, however, pinpoint the spatial position of the photon along the wavetrain because of the uncertainty principle discussed in the following section.

Photon energy is defined by an expression similar to Eq. (1.12),

$$\mathscr{E}(1 \text{ photon}) = h\nu = \frac{hc}{\lambda} = hc\bar{\nu} \qquad (1.14)$$

Thus radiant energy can be expressed by stating the number of photons at a particular wavelength. Substituting in Eq. (1.14) the values of c and h from Eqs. (1.2) and (1.13), we get

$$\mathscr{E}(1 \text{ photon}) = \frac{1.9863 \times 10^{-19}}{\lambda} \text{ J} \qquad (1.15)$$

where λ is in μm. The number of photons per joule thus is

$$1 \text{ J} = 5.0345 \times 10^{18}\lambda \text{ photons} \qquad (1.16)$$

Since $1 \text{ W} = 1 \text{ J sec}^{-1}$, we write

$$1 \text{ W} = 5.0345 \times 10^{18}\lambda \text{ photons sec}^{-1} \qquad (1.17)$$

which is a good value to remember.

Photons have the particlelike attributes of mass and momentum. Considering first the mass, Einstein's equation for the equivalence of mass and energy states that

$$\mathscr{E} = mc^2 \qquad (1.18)$$

where in this instance m is the mass of a photon. From this and Eq. (1.14) we have

$$\mathscr{E} = mc^2 = h\nu = \frac{hc}{\lambda} \qquad (1.19)$$

Since the product of mass and velocity is momentum p, Eq. (1.19) can be written as

$$p = mc = \frac{\mathscr{E}}{c} = \frac{h\nu}{c} = \frac{h}{\lambda} \qquad (1.20)$$

If we take literally the idea that a photon is a localized bunch of energy in free flight, we can regard the photon as a particle having energy $\mathscr{E}$, mass m, and momentum p. The uncertainty principle, however, sets limits to the accuracy to which $\mathscr{E}$ and p can be jointly known.

From relativity theory the mass of a particle moving at velocity v is given by

$$m(v) = \frac{m_0}{[1 - (v^2/c^2)]^{1/2}} \qquad (1.21)$$

where m_0 is the rest mass of the particle. Then the relativistic energy is

$$\mathscr{E} = \frac{m_0 v^2}{[1 - (v^2/c^2)]^{1/2}} \qquad (1.22)$$

However, the photon travels with the speed c of light. But the denominator in Eq. (1.22) vanishes for the condition $v = c$, so that, if $\mathscr{E}$ is to be finite, the rest mass m_0 must be equal to zero. Thus the energy of a photon is entirely kinetic. The mass of a photon is found by combining Eqs. (1.15) and (1.18) to give

$$m = \frac{1.9863 \times 10^{-19}}{\lambda c^2} \text{ kg} \qquad (1.23)$$

For the wavelength $0.55\mu m$, which marks the maximum sensitivity of the human eye, substitutions in (1.23) give the mass of a photon as 4.02×10^{-36} kg.

That a photon has momentum is revealed in the effect investigated by Compton (1923), where an X-ray photon of high energy in a directed beam collides with one of the outer electrons of an atom serving as a target. Such electrons are bound to the atomic nucleus by energies of only a few electron volts, while X-ray photons can have energies of several hundred electron volts according to Eq. (1.14). Following a collision with such a photon, the electron recoils, that is, it is knocked free of the atom and moves away at a certain angle with respect to the photon beam. The original direction of photon travel is altered by the collision, and the photon is said to be *scattered.* Also, the photon energy is reduced by just the amount imparted to the electron as recoil energy. The collision is elastic in the sense that, although momentum is exchanged, the total momentum is conserved. The Compton effect is a strong confirmation of quantum theory, and a lucid treatment will be found in Beiser (1967). A large scale example of photon momentum is seen in the appearance of a comet moving across the sky. The streaming of the comet tail away from the sun is a result of the force imparted to this tenuous material by the momenta of sunlight photons.

1.3.3. Uncertainty and Complementarity

A fundamental tenet of quantum theory is the *uncertainty principle,* often called the *principle of indeterminacy.* Put forth by Heisenberg (1930) the principle assumes various forms but in effect states that the values of certain paired quantities of atomic magnitude cannot be determined simultaneously and exactly. Because the principle implies definite limits to the knowability of the physical world, it bears strongly upon philosophy and, in particular, upon the problem of knowledge. Several of these considerations are discussed by Guillemin (1968) in his introductory account of quantum theory.

With respect to a photon or any particle of atomic magnitude, the principle asserts the impossibility of determining its position and momentum simultaneously, with perfect accuracy. The impossibility does not arise from any defect of knowledge or crudity of measurement apparatus, but seems to lie in the structure of nature. Denoting the uncertainty in measuring the position of a particle by Δx, and the uncertainty in measuring its momentum by Δp, the accuracy of simultaneous measurements of x and p is restricted by

$$\Delta x \, \Delta p \geq \frac{h}{2\pi} \qquad (1.24)$$

The principle does not set limits to the accuracy of measuring either quantity alone, for then Δx or Δp can be made as small as techniques permit. The trouble comes in trying to measure both at the same time. If the position of an electron, for example, is measured very accurately, then this measurement unavoidably introduces an uncertainty into the measurement of momentum. If this latter quantity is measured anew to better accuracy, then position becomes more uncertain, and so on. In adding to the information in one area, we increase the ignorance in another. Quoting Eddington

(1935): "The observer is like the comedian with an armful of parcels; each time he picks one up he drops another one." Several ingenious thought experiments have been devised in attempting to outwit the restriction Eq. (1.24) but to no avail. The principle always wins.

Energy and time also are paired uncertainties according to

$$\Delta \mathscr{E} \, \Delta t \geq \frac{h}{2\pi} \qquad (1.25)$$

Here $\Delta \mathscr{E}$ is the uncertainty in measuring the energy of a particle and Δt is the uncertainty in measuring the time at which it has the energy $\mathscr{E}$. Again, the principle does not set limits to the accuracy of measuring either quantity alone. Applied to absorption and emission, where wavelengths and hence energy levels can be determined to good accuracy, the principle means that the details of the processes cannot be followed to an equivalent accuracy in time. We can infer some time details but cannot always resolve them. In a sense the birth of light, by which all things are seen, cannot itself be observed.

A helpful adjunct of quantum theory is the *principle of complementarity* which reconciles the seemingly contradictory viewpoints that light is waves and light is particles. The contradiction is stated by Rossi (1965): "Experimental evidence shows that light is a physical entity simultaneously endowed with some of the properties that we ordinarily ascribe to particles, and with some of the properties that we ordinarily ascribe to waves." The term *ordinarily* denotes a crucial factor in that our ordinary experience is on a macroscopic scale. The concepts of waves and particles, originating in that experience, cannot be applied in the microscopic world with the same assurance and implied mutual exclusion that we employ in the everyday world. In no single experiment does light simultaneously act as both particles and waves. Neither do electrons, which for a long time were mainly regarded as particles. We cannot give a final description of either matter or radiant energy by means of a single idea or word.

Thus in attempting to depict the complete nature of light both the quantum model and the wave model have advantages and limitations. This suggests that the two models are not mutually exclusive in an overall sense but rather are complementary. Each model enables us to handle quite different aspects of the total phenomenon, which is a very primitive thing. Such considerations as these are embodied in the *principle of complementarity*, originated and explained at length by Bohr (1949). Viewpoints derived from the principle are used in this book.

1.4. MOLECULAR ORIGINS OF SPECTRA

The origins of absorption and emission lie in exchanges of energy between molecules and the electromagnetic fields that are always roundabout. In this section we obtain a qualitative view of molecular internal energy and transitions. We consider first the three forms of this energy and the spectral region associated with each. Absorptive and emissive transitions and the resulting spectra are then reviewed. The absorption coefficient is defined, and the processes that give spectral lines their shapes and widths are noted. Temperature and pressure effects on the line structure of a spectrum are shown.

1.4.1. Forms of Internal Energy

Three forms of internal energy are attributed to a molecule: rotational $\mathscr{E}_{rot}$, vibrational $\mathscr{E}_{vib}$, and electronic $\mathscr{E}_{el}$. The energy in each form is quantized according to the sense of Eq. (1.12), and the permitted values are determined by quantum rules and factors of molecular structure. For convenience any of these forms may be called *internal* or *excited state energy*. These forms are additional to the kinetic energy of molecular translation $\mathscr{E}_{tr}$, which plays an important role through the medium of molecular collisions. The translational energy is not quantized but exists in a continuous range of values. At any instant the total energy $\mathscr{E}_{tot}$ of a molecule is given by

$$\mathscr{E}_{tot} = \mathscr{E}_{rot} + \mathscr{E}_{vib} + \mathscr{E}_{el} + \mathscr{E}_{tr}$$

At this stage we consider only internal energy; the translational form is treated in Chapter 2. In absorption the molecule captures a photon, which ceases to exist, and the molecule thereby makes a transition to a higher level of internal energy. In emission the molecule releases a photon and makes a transition to a lower level. Such transitions are *radiative* and produce spectral lines. Another type of transition, called *nonradiative*, is caused by collisions and distributes the store of energy throughout the molecular population.

The principal features of the internal energy forms are:

Kinetic Energy of Rotation. This energy resides in the rotation of the molecule as a unit body. The two or more nuclei that constitute the molecule, and account for nearly all of its mass, are held together by valence binding forces. These are balanced by internal repulsion forces effective only at very small internuclear distances. The binding forces are provided by the outer electrons which are shared by the nuclei. Although the internuclear spacing in a typical gas molecule is little more than 0.1 nm, it is large compared to the diameters of the nuclei, which are "point masses." Thus the molecule has moments of inertia about certain axes and can rotate about each of these in the manner of a macroscopic body. Transitions between energy levels produce changes in the kinetic energy of rotation, hence in the angular velocity of rotation. Because of the small energies involved in such transitions, spectral lines of pure rotation appear only at extreme infrared and millimeter wavelengths.

Potential–Kinetic Energy of Vibration. The valence bond holding the nuclei together is not absolutely rigid but can be stretched and compressed slightly, producing changes of internuclear distance. This elastic bond, in conjunction with the masses of the nuclei, allows the nuclei to vibrate about their equilibrium positions, analogously to a spring and mass system of macroscopic size. Depending on the number and configuration of the nuclei, several modes of vibration are possible. Transitions between vibrational energy levels involve greater amounts of energy than do those of rotation, and the corresponding spectral lines occur in the infrared at wavelengths usually less than 15 μm. When a molecule makes a vibrational transition, it usually makes a rotational transition as well. The spectrum from a group of molecules then consists of rotational lines grouped on each side of the spectral location that corresponds to the vibrational transition. These spectra are *vibration–rotation bands,* shown broadly in Figure 1.1 and to high resolution in Figure 1.2*b.*

Potential Energy of Electron Arrangement. When the outer electron arrangement that provides the binding force has a stable or equilibrium configuration, the molecule is at the ground state or zero level of electronic energy. When the configuration is altered to an unstable one by acquisition of energy, as from a violent collision or absorption of an energetic photon, the molecule has potential energy in electronic form. Only small quanta of energy are required to rotate the molecule; larger quanta are needed to start it vibrating; still larger quanta are required to rearrange the electrons. If the rearranging occurs when the molecule is already vibrating at large amplitude, the valence bond may be stretched beyond recovery. The molecule then dissociates into its constituent atoms. When transitions occur simultaneously in all three forms of energy, numerous vibration–rotation bands known as an *electronic band system* appear in the visible and ultraviolet regions.

Absorption and emission, which require quantum theory for explanation, have to be distinguished from electromagnetic scattering, which is explainable by wave theory. Both processes are radiative and occur in atomic–molecular domains, but there is little further similarity. In absorbing radiant energy a molecule undergoes a transition to a higher state of internal energy, as discussed in the foregoing. In emitting radiant energy the molecule undergoes a transition to a lower state. Such transitions, highly selective with respect to wavelength, are the essence of absorption and emission and mean that these processes are discontinuous.

In contrast, electromagnetic scattering by a molecule occurs when a passing (primary) wave sets into oscillation the electric charges that constitute the molecule. The oscillating charges act as an electric dipole and radiate a secondary wave. Since the dipole oscillates synchronously with the primary, driving wave, the secondary wave has the same frequency, hence wavelength, as the primary wave. There is a fixed phase relation between the two waves. The scattering process is continuous and produces no net change in the internal energy of the molecule.

A third category of interaction between radiant energy and matter is *Raman scattering*. This occurs when certain species of molecules are irradiated with near-monochromatic light at frequency ν_0, well removed from the molecular vibrational frequency ν. It is found that the scattered light contains not only the frequency ν_0, as predicted by the scattering theory just discussed, but also small amounts of light at the sum and difference frequencies $\nu_0 + \nu$ and $\nu_0 - \nu$. The Raman effect produces vibrational (and rotational) spectra in the visible and infrared, even from species which ordinarily have no radiative activity. Raman scattering is not within the scope of this book. Its classical and quantum aspects, and interpretations of its spectra, are treated by Herzberg (1961). Long (1977) gives an unexcelled tutorial coverage in considerable detail, with many general and specialized references.

1.4.2. Radiative Transitions and Spectra

Radiative transitions require that the molecule be coupled to an electromagnetic field so that exchanges of energy can take place. In most cases the coupling is provided by the electric dipole moment of the molecule. This moment exists whenever the effective centers of the positive and negative charges of the molecule have a nonzero separation. Most species of infrared-active molecules, for example, H_2O, CO, and NO, have permanent electric dipole moments because of asymmetrical charge distributions. The N_2 and O_2 species have symmetrical charge distributions, hence they have no permanent electric dipole moments and no infrared activity. However, both species have weak magnetic dipole moments that permit radiative activity in the ultraviolet and to a slight extent in the visible.

For generality consider the absorption of a high-energy photon so that electronic energy is involved. The absorbed energy $\mathscr{E}_{ab}$ is apportioned among the three forms according to

$$\mathscr{E}_{ab} = (\mathscr{E}' - \mathscr{E}'')_{rot} + (\mathscr{E}' - \mathscr{E}'')_{vib} + (\mathscr{E}' - \mathscr{E}'')_{el} \qquad (1.26)$$

where $\mathscr{E}'$ is the upper energy level and $\mathscr{E}''$ is the lower. Absorption of an infrared photon usually does not involve enough energy to produce an electronic transition; the third term on the right-hand side of the equation is then zero. To be absorbed, the incident photon must have that amount of energy required by Eq. (1.26), hence it has a wavelength according to Eq. (1.14). In the matching of a quantum of radiant energy to a quantum of molecular energy a unique spectral line is created, and we see the selectivity of absorption.

For an individual molecule the transition is random in time or spontaneous. For a volume of gas, however, the transition rate follows a probability function known as the *Einstein coefficient of absorption*. Because the rotational energies are relatively small, many of the levels above the lowest are populated at terrestrial temperatures. In a group of many molecules this allows various combinations of a single vibrational transition and numerous rotational transitions at different levels to occur, according to Eq. (1.26). When the incident radiant energy is broadband, as for natural sources such as the sun and the sky, there is a continuous supply of matching photons for each value of $\mathscr{E}_{ab}$ that can be formed from Eq. (1.26), creating a vibration–rotation band whose center wavelength is fixed by the vibrational transition. The resulting spectral lines are close together and may merge because of overlapping widths, as shown in Figure 1.8 for two lines. The absorption then appears continuous but the line structure can be disclosed by a spectrometer of high resolving power, as in Figure 1.2b.

Emission is the converse of absorption and is governed by similar factors. When a photon is emitted, the total decrease of internal energy $\mathscr{E}_{em}$ is represented by

$$\mathscr{E}_{em} = (\mathscr{E}' - \mathscr{E}'')_{rot} + (\mathscr{E}' - \mathscr{E}'')_{vib} + (\mathscr{E}' - \mathscr{E}'')_{el} \qquad (1.27)$$

As in Eq. (1.26), $\mathscr{E}'$ and $\mathscr{E}''$ are the energy values of the upper and lower levels of the transition, with the third term usually zero for infrared emission. The emitted photon has that amount of energy required by Eq. (1.27) and a wavelength given by Eq. (1.14). A unique spectral line is created, and here is the selectivity of emission.

Disregarding stimulated emission, which makes lasers possible but occurs only to a limited extent in the atmosphere, an emissive transition is spontaneous. Each photon is emitted in a random direction and with a randomly oriented polarization so that the emission by a volume of gas is *isotropic* and *unpolarized*. The spectral lines often are closely spaced, similarly to those of absorption and for the same reasons, and are subject to the same broadening and overlapping. The transition rate in a volume of gas follows a probability function known as the *Einstein coefficient of spontaneous emission.* An energy state for which this probability is very small is a *metastable state,* and a molecule in this state must get rid of its excess energy by collisional exchanges.

The absorption of a photon having an energy value according to Eq. (1.26) usually is not followed by the direct emission of an equivalent photon according to Eq. (1.27). There are two reasons for this nonequivalence. The first one involves the condition of *local thermodynamic equilibrium* (LTE), in which molecules lose some of their absorbed energy by collisions before they emit. The collisional exchanges spread this energy throughout the volume of gas. When emissive transitions do ultimately occur, they start from levels below those to which they were initially raised by absorption. An overall result of the exchanges is that a single kinetic temperature characterizes the gas, and the spectral distribution of the emitted energy is a function of the gas temperature. This is true regardless of the spectral distribution of the absorbed energy, so that the radiant fluxes entering and leaving a volume of gas usually have different spectral distributions.

The second reason for the nonequivalence is concerned with *fluorescence.* This happens with certain types of material when absorptive transitions involving electronic energy are closely followed by emissive transitions before collisions carry away the excess energy. When the emissive transitions end at higher levels than those from which the absorptive transitions ensued, the molecules are emitting less energy than they just absorbed. From Eq. (1.14) this means that the lines of emission are at longer wavelengths than those of absorption. Such emission lines are known as *Stokes lines of fluorescence,* and familiar examples are seen with those materials whose absorption of ultraviolet light (black light) produces emission of visible light. In Raman scattering noted in the preceding section, one set of the spectral lines is at a shorter wavelength (higher frequency) than that of the exciting light. These are called *anti-Stokes lines* because they contravene Stokes' rule that, in the case of fluorescence, only wavelengths longer than those of the exciting light can be emitted.

1.4.3. Absorption Coefficients

The efficacy of absorption by a particular molecular species is expressed by its spectral absorption coefficient, considered at this point because it is basic to understanding line shapes and widths. This coefficient must be distinguished from the Einstein probability coefficient noted in the preceding section. The concept of absorption coefficient is explained most easily for a narrow spectral interval and small amounts of absorption.

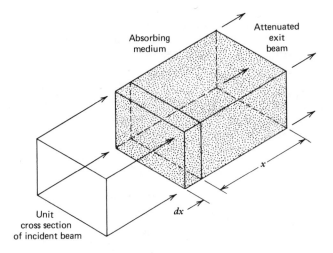

Figure 1.9. Attenuation of radiant flux in an absorbing medium. Each infinitesimal thickness of the medium removes a definite fraction of the incident flux.

Referring to Figure 1.9, let a beam of near-monochromatic flux, having spectral radiance L_λ be projected into a homogeneous absorbing medium. In path distance dx the fractional decrease of radiance dL_λ due to absorption is

$$\frac{dL_\lambda}{L_\lambda} = -k_\lambda \, dx \tag{1.28}$$

Here L_λ is the beam radiance incident at the lamina of thickness dx, and k_λ is a proportionality constant known as the *spectral absorption coefficient*. The value of k depends on the particular molecular species and concentration, and it usually is a strong function of wavelength.

Deleting for simplicity the subscript λ but remembering that we are dealing with spectral quantities, integration of Eq. (1.28) between the limits 0 and x gives

$$L_x = L_0 \exp\,(-kx) \tag{1.29}$$

where L_0 is the value at $x = 0$. This is Bouguer's law of attenuation, also called Lambert's law, and means that equal thicknesses of the medium will absorb equal fractional amounts of flux. The coefficient k is expressed as a numeral per unit of path length with dimension L^{-1}. Thus k defines the ability of unit path length in a specified medium to absorb flux at a given wavelength, hence it is a linear absorption coefficient. Actually k is associated with a volume because the idea of unit cross section is explicit in the definition of L, and it is often called a volume absorption coefficient.

The volume coefficient leads to a different but equivalent coefficient. In specifying a gaseous medium, a statement of its temperature and pressure defines its density ρ according to the gas laws. For such stipulations a unit volume represents a definite mass of gas, which is a more basic factor in absorption than is path length. The total mass u, per unit cross section, along a path of length x is simply

$$u = \rho x$$

with the dimensions ML^{-2}. Equation (1.29) then can be written

$$L_u = L_0 \exp(-Ku) \tag{1.30a}$$

where L_u is the beam irradiance at that path length which encompasses the total mass u of absorber. The term K is the mass absorption coefficient, having the dimensions $M^{-1}L^2$ and expressed as a numeral per unit mass (per unit cross section). In terms of transmittance τ, discussed in Appendix D, Eq. (1.30a) becomes

$$\tau = \frac{L_u}{L_0} \exp(-Ku) \tag{1.30b}$$

The linear and mass coefficients are related by

$$k = \rho K$$

Additional versions of the absorption coefficient, tied closely to molecular factors, are discussed in Section 6.2.3.

A good indicator of path absorptance is the *optical thickness* T, also called *optical depth*, defined by

$$T = kx \tag{1.31}$$

Equation (1.29) then becomes

$$L_x = L_0 \exp(-T) \tag{1.32}$$

which emphasizes that the attenuation depends on the product of linear coefficient and path length, but not on either factor alone. Similarly, the optical thickness in terms of the mass coefficient k is

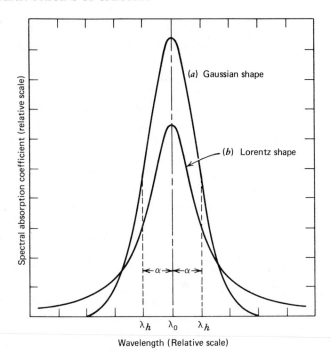

Figure 1.10. Spectral line shape produced by (*a*) Doppler broadening and (*b*) by collisional broadening. From Levi (1968).

$$T = Ku \tag{1.33}$$

and Eq. (1.30) becomes

$$L_x = L_0 \exp (-T) \tag{1.34}$$

with an equivalent interpretation.

1.4.4. Spectral Line Shapes and Widths

Although a spectral line of either absorption or emission covers only a small interval of wavelength, the shape and width, or profile, of the line are important. The profile derives from three causes: natural width, Doppler broadening, and collision broadening. Several common factors of shape and width are illustrated by the two curves in Figure 1.10, where the ordinate is the spectral absorption coefficient and the abscissa is wavelength.

The center wavelength of the line is that value at which the rapidly varying coefficient $k(\lambda)$ is a maximum. The half-width α is the distance from line center λ_0 to the wavelength λ_h where $k(\lambda)$ has decreased to one-half its maximum value. The value λ_h marks each of the half-power points. The two curves are drawn so that they have equal half-widths for comparison of shapes. Sometimes the unqualified term *line width* is encountered; this may mean α as defined here or it may mean 2α. The half-width is specified by its interval of wavelength, frequency, or wavenumber. The integral of $k(\nu)$ over the line shape is the line strength S.

The *natural width* of a line is due to the inherently brief duration of the emission, noted in Section 1.2.3. This duration is about 10^{-8} sec for an electronic transition, but is several orders of magnitude greater for rotational and vibrational transitions. The dependence of natural width on lifetime of the transition exemplifies that the frequency spectrum of a pulse becomes wider as the pulse becomes narrower. From the viewpoint of quantum mechanics the natural width is related by the uncertainty principle to the fact that the upper and lower energy levels of a transition are not perfectly discrete but are characterized by a kind of "fuzziness" or uncertainty of values. Natural width is only about 10^{-5} nm anywhere in the spectrum. This is too small to measure and can be ignored in comparison to the widths from the other causes. The natural line shape is called the *Lorentz shape,* after the Dutch physicist who investigated it following a suggestion from Michelson.

Doppler broadening is caused by the rapid translational motions of gas molecules, which in absorption are moving receivers and in emission are moving sources. The average speed of a CO molecule at 0°C, for example, is about 493 m sec^{-1} or 1103 mi hr^{-1}. The motions are completely random, and the *velocity* components along any direction of observation have a Gaussian distribution of magnitudes about the average value. Because one-half of the molecules have velocity components along a $+x$ direction and one-half have components along a $-x$ direction, the average velocity is zero according to this convention. The resulting Doppler-shifted wavelengths also have a Gaussian distribution about the center value λ_0, as shown by curve (a) in Figure 1.10. For example, the absorption line of CO at $\lambda = 4.67$ μm has a Doppler half-width $\alpha_D = 5.5 \times 10^{-1}$ nm at 0°C. Doppler broadening varies directly with gas temperature and is often called *temperature broadening.* An early calculation of Doppler broadening was made by Rayleigh (1889b).

Collision broadening is caused by disturbances to the absorbing and emitting processes during molecular collisions. These disturbances degrade the phase coherency of the *individual* process, and the resulting wavelength breadth of either an absorption or an emission line is greater than in the absence of a collision. The line shape, shown by curve (b) in Figure 1.10, is Lorentzian like the natural shape but the half-width α_L is several orders of magnitude greater. For a gas at 0°C and sea-level pressure, $\alpha_L = 4.22$ nm. Collision broadening varies directly with pressure and with the square root of temperature and the name *pressure broadening* is frequently used. The difference between the two line shapes in the figure is significant. The Gaussian shape is more peaked and compact, while the Lorentz shape has "wings" that extend far beyond the center wavelength. The resulting wing absorptions from lines in the H_2O and CO_2 bands adjacent to the atmospheric window at 8–13 μm, shown in Figure 1.1, produce measurable attenuation in this region.

The discreteness of spectral lines disappears as gas temperature and pressure are increased. At moderately high values of these parameters, the lines become sufficiently wide because of Doppler and collisional broadening that they overlap and begin to lose identity, as in Figure 1.8c. At still higher values some of the collisions are violent enough to cause molecular ionization and dissociation. The quantizations of internal energies now merge into a continuous range, producing a spectral continuum in which line structure has disappeared. An instance of this from a xenon flashtube is shown in Figure 1.11. The peak power in the electrical discharge was several megawatts, resulting in high temperature and pressure and ionization of the gas. Each of the two emission curves resembles that of a blackbody. At the lower current density the spectral structure of the emission is already coarse, appearing as broad bands superposed on a continuum. At the higher current density the structure is suppressed almost entirely, and the gas is emitting practically as a blackbody. Natural examples of gaseous sources radiating a blackbody continuum are solar and stellar atmospheres, where the temperatures are comparable to those indicated on the figure.

1.5. LAWS OF BLACKBODY RADIATION

Matter in its various thermal energy states is an important sink and source of radiant energy. The ideal sink and source is a *blackbody* whose absorptance and emissivity are each equal to unity at all wavelengths. A familiarity with blackbody principles is essential to an understanding of absorption and emission processes. In this section we review the main principles and laws of the blackbody for application to the atmosphere. First, the blackbody concept and Kirchhoff's law are discussed. The classical radiation laws and Planck's spectral distribution law are presented in terms of radiant power and photons. Additional relationships useful in

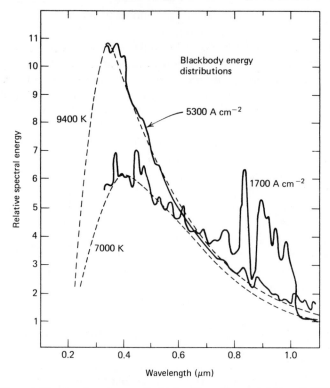

Figure 1.11.   Spectra of xenon flashtube. From Goncz and Newell (1966).

calculations are then given, along with references to tabulations of functions and calculator programs.

Readers may wish to pursue this area of study further than is possible here, and fortunately a large amount of literature is available. Good insights can be gained from the development of theory by Planck (1914). The treatment by Benford (1943a, b) is unsurpassed tutorially, while an abridged account is given by Benford (1939). An extended treatment by Bramson (1968) deals with many geometric factors and tabulated functions. Laboratory blackbody sources are described by Bramson (1968), Hudson (1969), Wolfe (1965), and Wolfe and Zissis (1979). Coverage emphasizing physics and thermodynamics is provided by Garbuny (1965), Jamieson et al. (1963), Kruse et al. (1963), Smith et al. (1957), and Williams and Becklund (1972). An alternative derivation of Planck's law was worked out by Einstein (1917), using the probabilities of absorptive and emissive transitions already noted. His procedure is discussed by Stone (1963) and Bramson (1968). In another type of derivation by Bose (1924), radiant

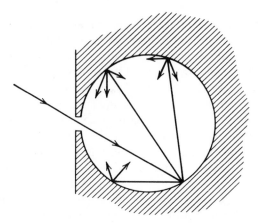

Figure 1.12. A blackbody radiation cavity

energy is regarded as a photon gas whose distribution in elementary cells of phase space is governed by Bose–Einstein statistics. The essentials of this method are reviewed by Beiser (1967) and Bramson (1968).

1.5.1. Blackbody Concept, Kirchhoff's Law

A blackbody is a basic concept of physics and easy to visualize. Consider a cavity fitted with a small port, as in Figure 1.12. Whatever the material and surface characteristics of the wall may be, most of the radiant flux entering the port from the outside is trapped within the cavity. As suggested by the ray arrows, repeated reflections occur until all the energy is absorbed by the wall. Only a small part of the entering flux is able to escape, so that the interior appears dark even when the port is illuminated from the outside. This is the reason for applying the name *blackbody* to a configuration of material for which the absorption is complete. The name derives from the visual impression.

Extrapolating from this model absorber, we realize that fairly complete absorption must occur in situations where a spatial configuration brings incident flux into repeated encounters with an absorbing material. At each encounter some of the original photons are absorbed until finally none remain. Such a result is obvious for a cavity, but it is no less true for a long optical path in an absorbing gas. Along a path from the earth's surface to the top of the atmosphere, water vapor, carbon dioxide, and ozone exhibit strong absorptions in numerous ultraviolet and infrared bands. When the geometrical path length is relatively great, the optical thickness defined by Eq. (1.33) then has large values in these bands. In such cases the gases absorb and emit practically as blackbodies. Finkelnberg (1947) called attention to these relationships in the early years of infrared technology.

Emission by a blackbody is the converse of absorption. As in Figure 1.12, the energy emitted by any small area of the wall is repeatedly reflected, and at each encounter with the wall the flux is weakened by absorption and strengthened by new emission. After many encounters for all rays the two processes come into balance. Emission reaches an equilibrium condition with respect to wall temperature, and the cavity contains a definable amount of steady-state radiant energy per unit volume. This quantity is the radiant density w, listed in Appendix C as one of the radiometric quantities. Just as the cavity absorbs energy at all wavelengths, so it emits energy at all wavelengths. An energy continuum exists. The value of radiant density and its spectral distribution are independent of cavity size and nature of the wall, and depend only on the cavity temperature. If the wall temperature is uniform and the small discontinuity of the port is ignored, the cavity energy is isotropic. Energy is being transported equally in all directions through any point, at the speed of light. At any instant one-half of this energy has components toward the plane of the port, from a hemisphere of directions. A definite amount of this energy escapes to the outside world so that the port has measurable values of spectral exitance M_λ and spectral radiance L_λ. This radiant flux is unpolarized and its spectral distribution follows Planck's law.

A basic relationship among the radiometric properties of matter and the laws of the blackbody is the principle called *Kirchhoff's law*. Briefly put (see Hudson, 1969), the law states that "at a given temperature the ratio of radiant exitance M_λ to absorptance α_λ is a constant and equal to the radiant exitance $M_{bb,\lambda}$ of a blackbody at that temperature." Thus we have

$$\frac{M_\lambda}{\alpha_\lambda} = M_{bb,\lambda} \qquad (1.35)$$

Combining the definition of emissivity ϵ in Appendix D with (1.35) we get an important identity

$$\epsilon_\lambda \equiv \alpha_\lambda \qquad (1.36)$$

for any material and for all α_λ from 0 to 1. This is a practical form of Kirchhoff's law, often stated as "good absorbers are good emitters."

In the following sections all the expressions for exitance and radiance, both spectral and total, are valid as written only for a blackbody where $\epsilon_{bb} = 1$. Most sources have emissivities less than unity, and these may vary with wavelength. Such a dependence is particularly true for atmospheric gases. Three types of sources are distinguished by the way in which the emissivity (hence the absorptance) varies:

A blackbody or *Planckian radiator* for which ϵ_{bb} is constant with wavelength and equal to unity.

A *graybody radiator* for which ϵ is constant with wavelength but is less than unity.

A *selective radiator* for which ϵ varies with wavelength, hence may be denoted by ϵ_λ.

Any of the expressions for exitance or radiance become valid for a graybody or a selective radiator when multiplied by the emissivity appropriate to the selected spectral region. Some useful relationships between emissivity and other radiometric properties of matter are summarized in Appendix D. Helpful discussions will be found in Hudson (1969), Jamieson et al. (1963), Kruse et al. (1963), and Sommerfeld (1964).

1.5.2. Classical Laws of Thermal Emission

During the latter part of the 19th century many studies were devoted to finding the relationships between the temperature of a blackbody and the total amount and spectral distribution of the emitted flux. One of the early results was the Stefan–Boltzmann law which expresses the total radiant exitance M,

$$M = \frac{2\pi^5 \kappa^4 T^4}{15\, c^2 h^3} = \sigma T^4 \ \text{W cm}^{-2} \tag{1.37}$$

where T denotes the temperature on the absolute scale and κ is Boltzmann's constant having the value

$$\kappa = 1.381 \times 10^{-23} \ \text{J K}^{-1} \tag{1.38}$$

and the dimensions ML^2T^{-2}. The symbol σ in Eq. (1.37) denotes the Stefan–Boltzmann constant,

$$\sigma = 5.6697 \times 10^{-12} \ \text{W cm}^{-2} \tag{1.39}$$

whose dimensions are MT^{-3}. The Stefan–Boltzmann law can be put in terms of radiance instead of exitance by means of the relation $M = \pi L$. This conversion applies to all the subsequent expressions for exitance, whether total or spectral.

The spectral distribution of blackbody radiant flux was a problem whose solution lay beyond classical physics. Before Planck discovered the exact law, two approximate solutions were devised. The first was proposed by Wien in 1896 as

$$M_\lambda = \frac{2\pi hc^2}{\lambda^5} \exp\left(-\frac{ch}{\lambda \kappa T}\right) \text{ W cm}^{-2} \text{ } \mu\text{m}^{-1} \qquad (1.40)$$

This is simplified by using two radiation constants c_1 and c_2 having the values

$$c_1 = 2\pi hc^2 = 3.7415 \times 10^4 \text{ W cm}^{-2} \text{ } \mu\text{m}^4 \qquad (1.41)$$

with dimensions ML^4T^{-3}, and

$$c_2 = \frac{hc}{\kappa} = 1.4388 \times 10^4 \text{ } \mu\text{m K} \qquad (1.42)$$

with dimension L. Equation (1.40) then becomes

$$M_\lambda = \frac{c_1}{\lambda^5} \exp\left(-\frac{c_2}{\lambda T}\right) \text{ W cm}^{-2} \text{ } \mu\text{m}^{-1} \qquad (1.43)$$

The Wien distribution agrees well with experimental data in the ultraviolet and visible, but poorly in the infrared.

The Wien distribution also showed the existence of a maximum for M_λ as a function of λ. This led to the Wien displacement law

$$\lambda_m T = 2897.8 \text{ } \mu\text{m K} \qquad (1.44)$$

where λ_m is the wavelength of maximum spectral exitance. Thus λ_m varies inversely with temperature, being 9.6 μm for an earth's surface temperature of 300 K, and near 0.55 μm for a nominal solar temperature (effective) of about 5500 K. This large spectral separation of the terrestrial and solar fluxes has great environmental importance.

The second approximate solution to the spectral distribution of blackbody flux is the Rayleigh–Jeans distribution,

$$M_\lambda = \frac{2c\kappa T}{\lambda^4} = \frac{c_1 T}{\pi c_2 \lambda^4} \text{ W cm}^{-2} \text{ m}^{-1} \qquad (1.45)$$

where the constants c_1 and c_2 are defined by Eqs. (1.41) and (1.42). This distribution agrees with experimental data only at very-long-wave infrared and, as may be seen from the λ^4 term, predicts an unlimited amount of flux

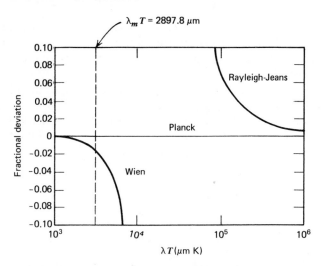

Figure 1.13. Fractional deviations of the Wien and Rayleigh—Jeans expressions from Plancks's exact law. From Jamieson et al., *Infrared Physics and Engineering.* © 1963 by McGraw-Hill, Inc. Used with permission of McGraw-Hill Book Co.

at short wavelengths. In the extreme infrared and microwave regions, this distribution has value because of its simplicity. The integral of Eq. (1.45) over a wavelength band λ_1 to λ_2 is

$$M_{\Delta\lambda} = \left| -\frac{c_1 T}{3 c_2 \lambda^3} \right|_{\lambda_1}^{\lambda_2} \text{ W cm}^{-2}$$

which is convenient for calculation.

Thus neither the Wien nor the Rayleigh–Jeans distribution is valid across the optical spectrum. The fractional deviations of these expressions from the Planck exact law, as functions of the product λT, are shown in Figure 1.13. The value $\lambda_m T = 2897.8$ μm K from Eq. (1.44) is plotted to show where the peak exitance occurs, regardless of the temperature. The Wien expression is accurate to 1% and better for $\lambda T < 3000$ μm K, and to better than 10% when $\lambda T < 6000$ μm K. The Rayleigh–Jeans expression is accurate to 1% and better when $\lambda T > 7.70 \times 10^5$ μm K ($= 7.70 \times 10^2$ mm K) and to better than 10% when $\lambda T > 7 \times 10^4$ μm K ($= 70$ mm K).

1.5.3. Planck's Spectral Distribution Law

Planck's expression for the spectral energy density w_ν of a blackbody, per unit frequency interval, is

$$w_\nu = \frac{8\pi h \nu^3}{c^3} \times \frac{1}{\exp(h\nu/\kappa T - 1)} \text{ J cm}^{-3} \text{ Hz}^{-1} \tag{1.46}$$

This leads to the expression for spectral exitance M_v by way of the following steps. Because the radiant energy traveling within the cavity is isotropic, one-half of it has velocity components toward the plane of the port from a hemisphere of directions. The time rate of radiant energy flow (radiant flux) across a unit area normal to the direction of propagation is given by the product of energy density and the velocity of light. Here we take the unit area to lie in the plane of the port. A multiplying factor of one-half must be applied because Lambert's cosine law governs the flux emitted at the plane of the port. Combining these ideas with Eq. (1.46), we see that the spectral exitance M_v of the port is given by

$$M_v = w_v \frac{c}{4} = \frac{2\pi h v^3}{c^2} \frac{1}{\exp{(hv/\kappa T - 1)}} \text{ W cm}^{-2} \text{ Hz}^{-1} \qquad (1.47)$$

Integration of Eq. (1.47) over all frequencies yields the Stefan–Boltzmann law, Eq. (1.37), as shown by Jamieson et al. (1963).

In terms of wavelength and the constants c_1 and c_2 defined by Eqs. (1.41) and (1.42), the spectral radiant exitance becomes

$$M_\lambda = \frac{c_1}{\lambda^5} \frac{1}{\exp{(c_2/\lambda T - 1)}} \text{ W cm}^{-2} \text{ } \mu\text{m}^{-1} \qquad (1.48)$$

per unit wavelength interval. This is a common form of Planck's law. When $\lambda T \ll c_2$ the exponential term becomes very large. The factor unity therein then can be ignored and Eq. (1.48) reduces to the Wien expression. Conversely, when $\lambda T \gg C_2$ the exponential term becomes very small and is approximated by the first two terms of its series expansion. Equation (1.48) then reduces to the Rayleigh–Jeans expression. The integral of Eq. (1.48) over all wavelengths is equal to the Stefan–Boltzmann law, Eq. (1.37).

Spectral exitance curves from Eq. (1.48) for several temperatures are plotted in Figure 1.14, while a larger family of spectral radiance curves is shown in Appendix E. The identical shapes of the curves are a consequence of log-log coordinates. Outstanding are the steady shift of the maxima toward shorter wavelengths, and the large increase of the exitance itself, as the temperature is increased. The dashed line linking the maxima becomes the trace of Wien's displacement law, Eq. (1.44). This function can be found directly from Eq. (1.48) by setting the derivative equal to zero and solving.

The value of spectral exitance at the wavelength of the maximum is found by substituting the value of λ_m from Eq. (1.44) into Eq. (1.48). We get

$$M_{\lambda m} = 1.2862 \times 10^{-15} \text{ } T^5 \text{ W cm}^{-2} \text{ } \mu\text{m}^{-1} \qquad (1.49)$$

showing that the spectral exitance at the maximum varies as the fifth power

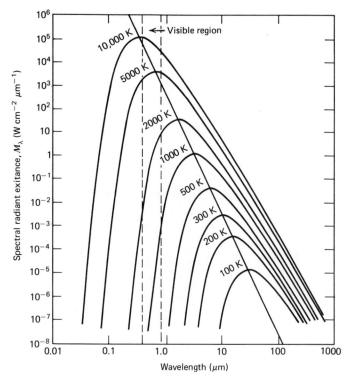

Figure 1.14. Spectral radiant exitance according to Planck's exact law.

of temperature. Thus the ratio of two values of exitance at their maxima equals the ratio of their temperatures to the fifth power. As an example, merely doubling the temperature produces a 32-fold increase of M_λ at λ_m, which may be seen in Figure 1.14. However, λ_m has quite different values in the two cases.

1.5.4. Blackbody Laws for Photons

The processes of photoelectric detection, photochemical reactions, and absorption–emission depend on the rate at which photons are involved. Worthing (1939a, b) showed that to each radiation law there corresponds a photon law, and he provided a double-entry table listing the radiation constants in terms of radiant power and in terms of photons. In this section we state in terms of photons per second three major radiation laws: (1) the Planck spectral distribution, Eq. (1.48), (2) the Stefan–Boltzmann total exitance, Eq. (1.37), and (3) the Wien displacement function, Eq. (1.44).

Dividing the Planck expression, Eq. (1.48), by hc/λ, which from Eq. (1.14) is the energy of one photon, we find for the spectral photon exitance

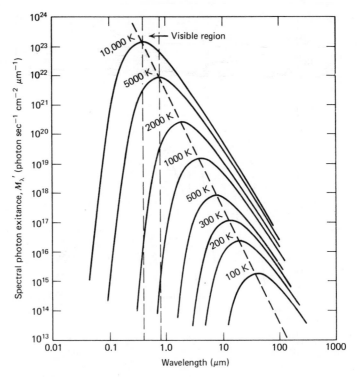

Figure 1.15. Spectral photon exitance according to Planck's exact law.

$$M'_\lambda = \frac{2\pi c}{\lambda^4} \times \frac{1}{\exp{(hc/\lambda\kappa T - 1)}} \text{ photon sec}^{-1} \text{ cm}^{-2} \text{ } \mu\text{m}^{-1} \quad (1.50)$$

where the prime superscript signifies a photon quantity. Using the second radiation constant c_2 defined by Eq. (1.42), and a new constant c'_1 defined by

$$c'_1 = 2\pi c = 1.88365 \times 10^{23} \text{ sec}^{-1} \text{ cm}^{-2} \text{ } \mu\text{m}^3 \quad (1.51)$$

we rewrite Eq. (1.50) as

$$M'_\lambda = \frac{c'_1}{\lambda^4} \times \frac{1}{\exp{(c_2/\lambda T - 1)}} \text{ photon sec}^{-1} \text{ cm}^{-2} \text{ } \mu\text{m}^{-1} \quad (1.52)$$

When the Planck law is thus expressed in terms of photons, wavelength appears to the fourth power, not the fifth. Figure 1.15 shows a plot of Eq. (1.52) for the same temperatures as in Figure 1.14. A larger family of photon spectral radiance curves appears in Appendix E.

The Stefan–Boltzmann law Eq. (1.37), is obtained in terms of photons by integrating Eq. (1.52) from $\lambda = 0$ to $\lambda = \infty$. Some details of the integration are given by Benford (1943b). The result is

$$M' = \sigma' T^3 \text{ photon sec}^{-1} \text{ cm}^{-2} \qquad (1.53)$$

where the constant σ', corresponding to the original Stefan–Boltzmann constant σ defined by Eq. (1.39), has the value

$$\sigma' = 1.5204 \times 10^{11} \text{ sec}^{-1} \text{ cm}^{-2} \text{ K}^{-3} \qquad (1.54)$$

Thus the total number of photons emitted per second varies as the third power of temperature.

This displacement law, Eq. (1.44), when written for photons gives the wavelength λ_m at which the maximum rate of photon emission occurs. This is found by setting the derivative $dM_\lambda'/d\lambda$ of Eq. (1.52) equal to zero and solving for $\lambda_m' T$. This gives

$$\lambda_m' T = 3669.7 \ \mu\text{m K} \qquad (1.55)$$

The spectral exitance in photons per second at λ_m is the counterpart of Eq. (1.49) and is found by substituting the value of λ_m' from Eq. (1.55) into Eq. (1.52). We find

$$M_{\lambda m} = 2.1010 \times 10^7 T^4 \text{ photon sec}^{-1} \text{ cm}^{-2} \ \mu\text{m}^{-1} \qquad (1.56)$$

The dependence on temperature to the fourth power, not the fifth as in Eq. (1.49), should be noted.

1.5.5. Aids to Calculation

All plots of Planck's law on log-log coordinates have the same shape and can be made to coincide by shifting the curves. Also, the Wien displacement law becomes a straight line on this type of plot. Moon (1936) has used these properties to devise an analog method of producing exitance and radiance curves. This requires a log-log framework with exitance or radiance units as the ordinate scale and wavelength units as the abscissa scale. The displacement law is then plotted and graduated over its length with a logarithmic temperature scale. Next a transparent overlay or template of a single curve—say the one for 10,000 K in Figure 1.14—is prepared. The overlay or template is then slid along the graduated line of displacement until the peak of the curve coincides with the selected temperature, while maintaining orientation with the coordinate grid. Values of exitance or

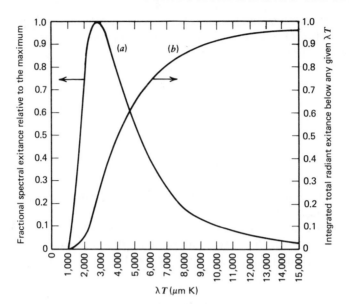

Figure 1.16. Blackbody universal curves for (*a*) fractional spectral radiant exitance and (*b*) integrated total radiant exitance. From Hudson (1969).

radiance versus wavelength now can be read from the intersections of the curve with the ordinate and abscissa scales. The curves in Figures 1.14 and 1.15 were drawn in this manner from a template.

The full meaning of the Wien displacement law is not limited to showing the wavelength where the maximum exitance occurs. A further consequence of the law is that, for all temperatures, the ratio of the spectral exitance at a given wavelength to the spectral exitance at λ_m is independent of the temperature. This fact allows a universal exitance curve to be constructed for all values of the product λT, as shown by curve (*a*) in Figure 1.16. At any value of λT the left ordinate gives the fractional exitance relative to the maximum at $\lambda T = 2897.8$ μm K. This maximum can be calculated from Eq. (1.49) or read from tabulations of this function. Multiplication of this absolute value by the fractional value just obtained from Figure 1.16 for the selected λT gives the absolute value of the exitance.

The curve (*b*) in Figure 1.16 shows the integrated exitance between zero and any selected value of λT as a fraction of the total exitance defined by the Stefan–Boltzmann law, Eq. (1.37). This fractional exitance, whose units are the right ordinate scale of the figure, is the value of the ratio $M_{0,\lambda}/M_{0,\infty}$. The numerator is the integral of Eq. (1.48) over wavelengths from zero to the selected value, and the denominator is the integral over wavelengths from zero to infinity. This denominator represents, of course, the Stefan–Boltzmann law. Many tabulations of this ratio appear in the literature; the one from List (1966) is reproduced in Appendix F. We also note that the fractional exitance over a wavelength band is equal to the difference of the two fractional amounts corresponding to the long-wave limit λ_2 and the short-wave limit λ_1 of the band. Hence the absolute value of exitance for a given band and temperature is determined by (1) taking the difference between values for $\lambda_2 T$ and $\lambda_1 T$ from curve (b) or Appendix F, and (2) multiplying this difference by the value of $M_{0,\infty}$ as found from Eq. (1.37) or tabulations of this function.

Blackbody literature provides many tabulations of functions covering wide ranges of temperature and wavelength. Those cited below are sufficiently common that readers will find one or more of them at hand. The radiometric symbols in this list are SI symbols, which are not always the same as those used in the cited works. The explanations accompanying the tabulations, however, define the quantities.

Gray (1957): $M_{0,\lambda}/M_{0,\infty}$ and $M_{\lambda}/M_{\lambda,m}$ for $\lambda T = 0.05$–2.0 cm K, and $M_{0,\infty}$ and $M_{\lambda,m}$ for $T = 1$–10,000 K.

List (1966): $M_{0,\lambda}/M_{0,\infty}$ for $\lambda T = 0.05$–2.0 cm K. This tabulation is reproduced in Appendix F.

Levi (1968): $M_{0,\lambda}/M_{0,\infty}$ and $M_{\lambda}/M_{\lambda,m}$ for $\lambda T = 300$–9900 μm K, and $M_{\lambda,m}$ and M_{λ} for $T = 100$–9900 K. Also given is λ_m for $T = 200$–50,900 K. These and additional tabulations will be found in Levi (1974).

Pivovonsky and Nagel (1961): The most extensive tabulations in the literature. Included are L_λ and $L_{0,\lambda}/L_{0,\infty}$ for separate arguments of wavelength and temperature. The total range of wavelengths is 0.20–1100 μm; the total range of temperatures is 20–40,000 K.

Bramson (1968): M, $M_{\lambda,m}$ and λ_m for the temperature range 90–6000 K, and M_λ and $M_{0,\infty}$ for the wavelength range 0.9–15.0 μm and the temperature range 300–2000 K. Also tabulated for these same ranges are the photon quantities M_λ' and M'.

For many purposes the values of blackbody functions can be obtained more conveniently with programmable electronic calculators than from tables. Both Texas Instruments (TI) and Hewlett-Packard (HP) can supply programs for certain models of their calculators that allow functions to be computed quickly to a high level of accuracy. For the TI calculator models SR52, TI58, and TI59, programs have been written by Dix (1976), McFee (1976a, b), and TI (1978). For the HP calculator models HP41c, HP67, and HP97, available blackbody programs are identified under HP (1982).

2

Gas Properties,
Thermodynamics,
Molecular Kinetics

A throng of tiny bodies in the void,
... with relentless movement driven
In paths diverse, some, when together dashed,
Leap back great space apart, while some are thrust
But short way from the blow ... [Lucretius (c60 BC)]

In *The Restless Universe* Max Born (1951) remarks that there is no condition in the physical world corresponding to the familiar idea of *rest*. The atoms and molecules of every substance are in ceaseless motion, which is the essence of all thermal energy. In liquids and solids the atoms–molecules congregate to a closely involved community where they are constrained by dense packing and interaction to vibratory chattering about average positions. Gas molecules are loners and wanderers. Having no fixed residences, they move randomly and rapidly from one collision to another, rotating and vibrating as they go, like a troop of circus tumblers swarming across an arena. In these total molecular activities, and the resulting exchanges of energy with electromagnetic fields, lie the basic phenomena of absorption and emission.

This chapter begins by reviewing the concepts employed to define and measure the macroscopic properties of gases. The properties considered are temperature, pressure, volume, and density, in an atmospheric context. The characteristics of water vapor are reviewed, along with methods of specifying its amount. Molar quantities and the gas laws are discussed, and the equation of state is developed. Thermodynamic states and laws are summarized and attention is given to entropy. All of these properties form a base from which to view the molecular world revealed by the kinetic theory of gases. The idea of a perfect gas is described; elastic and inelastic collisions are distinguished; and molecular quantities are reviewed. Maxwell–Boltzmann distributions of molecular populations over the ranges of velocity, speed, and energy are treated. The chapter concludes with a discussion of local thermodynamic equilibrium and its role in absorption–emission.

2.1. GAS PROPERTIES AND MEASUREMENT UNITS

The era of science began about 1600 when speculations dating to Aristotle slowly gave way to rational theorizing and related measurement. For two centuries and more studies of gas properties were a major concern of physics, persisting through the development periods of gas laws, kinetic theory, thermodynamics, and statistical mechanics. The very name *gas* suggests something light and insubstantial, and to the early experimenters the properties must have seemed as elusive as this intangible substance itself. In his diary of 1663 Samuel Pepys tells that King Charles II, even though a patron of the newly formed Royal Society, "mightily laughed at its members for spending time only in the weighing of ayre." The fact that gas can be compressed was a significant discovery by Boyle, who referred to this property as "the spring of the air." Because of its molecular motions, gas expands at every opportunity to fill the space around it, in the manner that Parkinson's law prescribes for bureaucratic activity. At this point we review the properties and measures of gases. Special attention is given to water vapor because of its unique characteristics and importance to absorption–emission.

2.1.1. Temperature, Pressure, Associated Quantities

Temperature is an important property of matter because it is the measure of the thermal energy called *heat* inherent in every substance. In physics temperature is usually measured on the Kelvin (K) or absolute scale. Referring to Figure 2.1, the 0 K condition is that of *absolute* zero, where thermal energy is almost completely absent. The size of a Kelvin degree is defined by the statement that the *triple point* of water occurs at 273.16 K. This is the temperature at which the three phases—solid, liquid, and vapor—are in equilibrium. The ice point of water is at 273.15 K. In meteorology temperature is measured on the Celsius scale, wherein the ice point of water occurs at 0°C and the boiling point at 100°C. As seen in Figure 2.1, the size of a Celsius degree is the same as that of a Kelvin degree.

The third temperature scale in the figure was devised by Fahrenheit, physicist of Danzig. Because a mixture of salt and ice produces a readily attainable low temperature, he set the zero of his scale at that point. For an upper point he proposed (at first) the blood temperature of a healthy man, which he fixed at 96 degrees, somewhat arbitrarily. Subsequently he selected for an upper point the temperature of boiling water, which turned out to be 212 degrees on his scale. The freezing point occurred at 32 degrees, so that 180 degrees covers the span from freezing to boiling. Thus we have for conversions:

Degrees Fahrenheit = (1.8 × degrees Celsius) + 32

Degrees Celsius = (degrees Fahrenheit − 32) × 0.56

Temperature is denoted generally by T, particularly on the Kelvin scale, and is a fundamental unit in the SI system.

Pressure P is defined as force per unit area, hence has dimensions

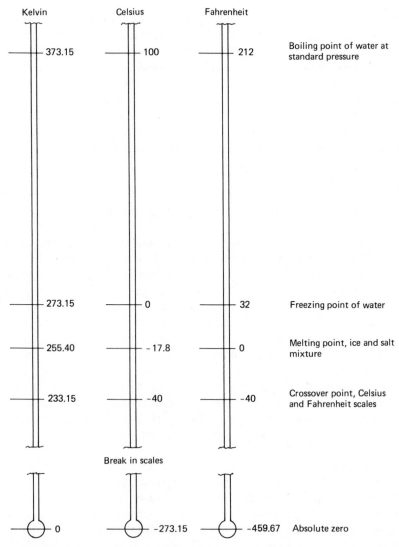

Figure 2.1.   Relationships between Kelvin, Celsius, and Fahrenheit temperature scales.

$ML^{-1}T^{-2}$, and several measurement units are in use. The nominal (average) pressure exerted by the atmosphere at sea level is taken as equal to the base pressure of a mercury column 760 mm high at 0°C. This is equal to 1.01325×10^6 dyn cm^{-2}, which is referred to as *one atmosphere* (1 *atm*) of pressure, or *standard pressure*. A frequently used unit, especially in the laboratory, is the *torr,* which is named for Torricelli who invented the mercury barometer in 1643. The torr and the mm Hg are identical. Another measure of pressure called the *bar* (used but little) has the value 10^6 dyn cm^{-2}. In meteorology the customary unit is the *millibar* (mb), which is equal to 10^3 dyn cm^{-2}. Therefore, standard pressure of 760 mm Hg, or 1.01325×10^6 dyn cm^{-2}, equals 1013.25 mb. In the SI and MKSA systems the unit of force is the *newton* (N), equal to 10^5 dyn. Thus standard pressure becomes 1.01325×10^5 N m^{-2}. When gas is at 0°C and 1.01325 mb it is said to be at standard temperature and pressure (STP).

In the SI system a force of 1 N m^{-2} is called the pascal (Pa) in honor of the French philosopher and scientist. It was he who first carried Torricelli's barometer to the top of a tower and showed that atmospheric pressure decreases as altitude increases. Standard pressure in this SI unit is 1.01325×10^5 Pa. Summarizing several of the above measures of atmospheric pressure:

$$1 \text{ atm} = 760 \text{ mm Hg} = 760 \text{ torr} = 1.01325 \times 10^6 \text{ dyn cm}^{-2}$$
$$= 1.01325 \times 10^5 \text{ N m}^{-2}$$
$$= 1.01325 \times 10^5 \text{ Pa}$$
$$= 1013.25 \text{ mb} \qquad (2.1)$$

The concept of volume V, having dimensions L^3, deserves consideration when applied to atmospheric air. In a laboratory the idea of volume usually refers to the space enclosed by a container whose size can be adjusted to vary the volume. The situation is otherwise in the free atmosphere, where any part is free to move—translate, expand, or contract—in response to forces exerted by gravity and adjacent masses of air. Here it may be helpful to think of *parcels* of air that are enclosed by imaginary interfaces something like elastic membranes. Closely related to volume is density ρ, defined as mass per unit volume and having dimensions ML^{-3}. The term *specific volume* refers to the volume occupied by unit mass of gas at stated temperature and pressure. Specified volume is defined by

$$V = \frac{1}{\rho} \qquad (2.2)$$

and has dimensions $M^{-1}L^3$.

When a gas is compressed its temperature increases, and when it is allowed to expand by doing work, its temperature decreases. If the compression or expansion occurs without transfer of heat between the gas and its surroundings, the process is *adiabatic*. Because air is a rather poor conductor of heat many atmospheric processes are approximately adiabatic. A familiar example of adiabatic compression is a wind blowing down a mountain slope to lower altitudes. The resulting compression increases the temperature of the moving air, producing a *foehn* wind. Such warm, dry winds on the lee slopes of mountains are common in many regions. In the Pacific Northwest these winds are called *chinooks* and often cause spectacular melting of snow fields within a few hours. Daily examples of adiabatic expansion are seen in the forming of fair-weather cumulus clouds. Convective updrafts carry the surface air to higher altitudes, where the resulting expansion cools the air to the condensation value for the entrained water vapor. Condensation then occurs on the particles and nuclei always present in the atmosphere.

The rate of temperature change with altitude Z in the atmosphere is the *lapse rate* γ, defined by

$$\gamma = -\frac{dT}{dZ} \tag{2.3}$$

where the minus sign makes γ positive for the usual decrease of temperature with altitude. Two types of lapse rate are distinguished. When the temperature change is associated with a rising or subsiding flow of air, we speak of the *process lapse rate*. For an adiabatic process it has the value

$$\gamma = 0.98°C \text{ per } 100 \text{ m} \quad \text{(process)} \tag{2.3a}$$

A derivation of this value is given by Fermi (1937). The second type is the *environmental lapse rate*, which is the average due to all meteorological processes. Its value is

$$\gamma = 0.65°C \text{ per } 100 \text{ m} \quad \text{(environmental)} \tag{2.3b}$$

in the lowest 10 km or so of the atmosphere, the region known as the *troposphere*. At higher altitudes, as in the stratosphere, mesosphere, ionosphere, and above, the lapse rate has other values, being variously zero, negative, or positive.

2.1.2. Measures of Water Vapor

Water vapor is the most active meteorological agent of all the atmospheric gases. Its supply is unlimited because more than three-fourths of the earth's surface is covered by water. Diffusing into the air by evaporation and transported great distances by winds and convection currents, water vapor is a principal absorber of solar and terrestrial radiant energy and is a moving reservoir of energy for meteorological processes. The ability of water to store and release energy is indicated by its unique thermal characteristics:

$$\text{Specific heat} = 1 \text{ cal g}^{-1}\,{}^{\circ}\text{C}^{-1} = 4.1855 \text{ J g}^{-1}\text{ K}^{-1}$$

$$\text{Heat of fusion} = 80 \text{ cal g}^{-1} = 334.84 \text{ J g}^{-1}$$

$$\text{Heat of vaporization} = 540 \text{ cal g}^{-1} = 2260.17 \text{ J g}^{-1}$$

The amount of water vapor that can be evaporated into a given volume of air depends only on the air temperature. This amount is defined either by the vapor density ρ_v or by the vapor pressure e_v, which is a partial pressure according to Dalton's law. The saturation vapor pressure is the maximum pressure that can be produced at a given temperature by evaporation from the liquid or solid phase of water. When this pressure is reached, the rate of molecular entry into the gas phase is just equalled by the rate of molecular return to the first phase. The two phases are then in equilibrium, and the space containing the vapor is saturated. A similar meaning attaches to saturation density. Values of ρ_v and e_v as a function of T are plotted in Figure 2.2. The rapidly increasing capacity of air to contain water vapor as the temperature is raised should be noted. If the air is originally saturated at a given temperature and the temperature is then reduced, the air becomes supersaturated. This is the usual condition within a cloud.

The amount of water vapor in a volume of air is expressed in several ways. Absolute humidity is the ratio of the mass of water vapor present to the volume of air. This quantity is defined by the vapor density and the usual units are g m^{-3}. Relative humidity, at a stated temperature, is the dimensionless ratio of actual (partial) vapor pressure (or vapor density) to the saturation value for that temperature and is expressed in percent. Mixing ratio is the mass of water vapor contained in unit mass of dry air; hence, it is equal to the ratio of the two densities.

Dewpoint T_d and frostpoint T_f are often employed as measures of water vapor, particularly in atmospheric soundings. Dewpoint is the temperature

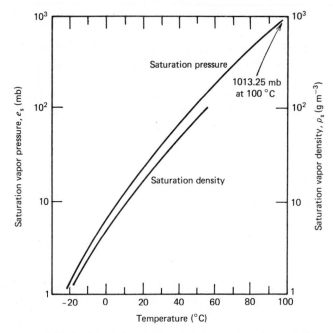

Figure 2.2. Saturation pressure and saturation density of water vapor, over a plane surface of pure water, as functions of temperature. Data from List (1966).

to which a parcel of air must be cooled, at constant pressure and constant vapor content, for saturation and resultant condensation to occur; that is, dewpoint is the temperature at which the partial vapor pressure would be the saturation pressure. Frostpoint temperature is defined the same way except that the saturation pressure is with respect to ice, so that the condensation produces frost. Numerical relations between the foregoing measures of water vapor are listed in Table 2.1.

The amount of water vapor along a path is specified by the depth of liquid water that would be produced if all the vapor were precipitated. Precipitable water in centimeters (pw cm) is the counterpart of the unit *atm cm* used for gases that do not condense in the atmosphere. Since the density of water is 1 g cm^{-3}, the precipitable water per kilometer of path length is given by

$$pw\ cm\ km^{-1} = 0.1\rho_v \qquad (2.4)$$

Here ρ_v is the average vapor density in g m^{-3} and the factor 0.1 reconciles the units. Figure 2.3 shows pw cm per kilometer and per nautical mile as functions of temperature and relative humidity. The chart is applicable to a homogeneous path or to one where a valid mean temperature and relative humidity can be assigned.

2.1.3. Molar Relationships

When dealing with the properties of gases it is often convenient to employ the quantities *molar mass* and *molar volume*. Recall that a gram molecular weight, or mole, of a substance is a mass in grams numerically equal to the molecular weight of that substance. Thus a mole represents a definite amount of substance called a *molar mass* M_A, where the subscript is a reminder of the early Italian chemist Avogadro. Evidently, the amount of substance in a mole, or molar mass, does not depend on the temperature or pressure of the substance. The mole is a base unit of the SI system, as listed in Appendix B. The volume occupied by a mole of gas is the *molar volume* V_A and at STP conditions has the value

$$V_A = 2.242 \times 10^4 \text{ cm}^3 = 22.42 \text{ liter}$$
$$= 2.242 \times 10^{-2} \text{ m}^3 \qquad (2.5)$$

TABLE 2.1. CONVERSION OF DEWPOINT TO OTHER MEASURES OF MOISTURE

Dewpoint, T_d (°C)	Frostpoint, T_f (°C)	Vapor Pressure, e_v (mb)	Vapor Density, ρ_v (g m^{-3})
−40	−36.5	0.189	0.176
−35	−31.8	0.314	0.286
−30	−27.2	0.509	0.453
−25	−22.5	0.807	0.705
−20	−18.0	1.254	1.074
−15	−13.4	1.912	1.605
−10	−8.9	2.863	2.358
−5	−4.5	4.215	3.407
0	0	6.108	4.847
5		8.719	6.797
10		12.272	9.399
15		17.044	12.83
20		23.373	17.30
25		31.671	23.05
30		42.430	30.38
35		56.236	39.63
40		73.777	51.19

Source: Gringorten et al. (1966).

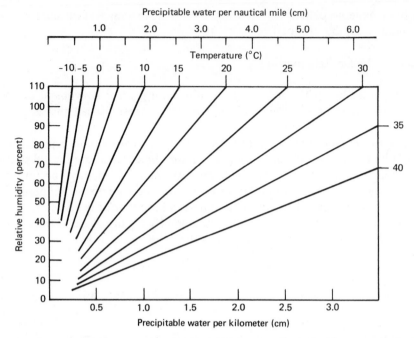

Figure 2.3. Precipitable water as a function of temperature and relative humidity. From McCartney (1976).

Avogadro's law, offered as a hypothesis in 1811 but abundantly verified since, states that "Equal volumes of gas, under the same conditions of temperature and pressure contain equal numbers of molecules." The number of molecules in 1 molar mass, or 1 mole, of any gas, regardless of temperature and pressure, is Avogadro's number,

$$N_A = 6.025 \times 10^{23} \text{ mole}^{-1} \tag{2.6}$$

The number of molecules in 1 cm³ of any gas at STP conditions is Loschmidt's number,

$$N_L = 2.687 \times 10^{19} \text{ cm}^{-3} \tag{2.7}$$

The number of molecules in any unit volume under consideration, with the temperature, pressure, or mass density specified when necessary, is the molecular number density N.

Because air is a mixture of gases, there is no unique species of "air molecule." Nevertheless, it is convenient to deal with a fictitious air molecule whose molecular weight is a weighted average of the three principal atmospheric gases—nitrogen, oxygen, and argon. These are present in volume percentages 78.1, 20.9, and 0.9, respectively. Application of these percentages to the molecular weights listed in Table 1.1 yields a molecular weight 28.964 (dimensionless) for the air molecule. Hence a molar mass of air has the value 28.964 g regardless of temperature and pressure.

The mass m of a single molecule is equal to the molar mass divided by Avogadro's number, or

$$m = \frac{M_A}{N_A} \tag{2.8}$$

substitutions therein give for the mass of an air molecule,

$$m = \frac{28.964}{6.025 \times 10^{23}} = 4.81 \times 10^{-23} \text{ g}$$

$$= 4.81 \times 10^{-26} \text{ kg} \tag{2.8}$$

The specific volumes of 1 g and 1 kg of air are

$$V = \frac{V_A}{M_A} = 7.74 \times 10^2 \text{ cm}^3 \text{ g}^{-1}$$

$$= 0.774 \text{ m}^3 \text{ kg}^{-1} \tag{2.9}$$

In general terms the mass density of a gas is equal to the product of molecular mass and number density,

$$\rho = mN = \frac{M_A}{V_A} \tag{2.10}$$

The mass density of STP air is then given by

$$\rho = mN_L = 1.29 \times 10^{-3} \text{ g cm}^{-3}$$

$$= 1.29 \text{ kg m}^{-3} \tag{2.10a}$$

2.2. GAS LAWS AND THERMODYNAMICS

The observable behavior of gases is described by the gas laws in terms of gross properties such as temperature, pressure, and density. Thermodynamics carries the behavior description further and, in so doing, establishes a bedrock of physics. It deals with the heat (energy) and temperature of a specified system and with conversions between heat and other forms of energy. Here we restrict the term *system* to mean an assemblage of gas molecules isolated from their surroundings. However, the classical laws of gases and thermodynamics do not convey any microscopic explanation of how the system really works, but rather are broad generalizations based on experiment. A microscopic view is provided by the kinetic theory of gases, discussed in Section 2.3, which applies the laws of mechanics to individual molecules. The method of statistical mechanics, treated in Chapter 3, forsakes the individual and employs statistical methods to handle the large numbers of molecules and energy states that make up a macroscopic system. Thermodynamics complements both kinetic theory and statistical mechanics, and the summary presentation of its principles here provides a basis for later discussions of these subjects.

In this section we review the gas laws of Boyle and Charles. By examining the rationale for the absolute scale of temperature we arrive at the concept of the universal gas constant and the equation of state for an ideal gas. Thermodynamic principles and the concept *entropy* are discussed in the context of a gas system. Our treatments of such pervasive subjects are necessarily brief and many readers will wish to inquire further. Good texts are Kittel (1969), Kittel and Kroemer (1980), Sears and Salinger (1975), van Wylen and Sonntag (1976), and Zemansky (1957). The presentations by Jancovici (1973) and Wäser (1966) are very clear and very concise. Sommerfeld (1964) treats all aspects of gas laws and thermodynamics with his usual elegance.

2.2.1. Laws of Ideal Gases

The gas laws in elementary form refer to an ideal or perfect gas. The term *ideal* involves the assumption that the molecules exert forces on each other only during collisions. In effect this says that the average spacing of the molecules is much greater than the range of intermolecular forces of attraction. Hence the van der Waal's correction, which takes into account this attraction, need not be used. As shown later, a gas at STP conditions is not quite "ideal," but is sufficiently so that introductory treatments usually make this assumption.

First we consider Dalton's *law of partial pressures,* which states that the total pressure exerted by a mixture of nonreacting gases is equal to the sum of the partial pressures of the component gases. Each partial pressure is that which each gas alone would exert in the same volume at the same temperature. For example, if components 1, 2, 3 exert partial pressures P_1, P_2, P_3, the total pressure exerted by the mixture is

$$P_{total} = P_1 + P_2 + P_3 \tag{2.11}$$

In 1662 the experiments of Boyle, and those by Mariotte independently, showed that the relation between pressure and volume of a gas, when temperature is held constant, is given by

$$PV = \text{constant} \tag{2.12}$$

from which we see the aptness of Boyle's analogy "the spring of the air." The relationship (2.12) is the Boyle–Mariotte law, and the meaning and value of the constant are explained in the following.

Near the end of the 18th century, Gay-Lussac and Charles investigated the behavior of gases at constant volume and constant pressure while the temperature was varied. When volume was held constant, pressure varied with temperature according to

$$P = P_0(1 + At) \tag{2.13a}$$

where P is the pressure at temperature t on the Celsius scale, P_0 is the pressure at 0°C, and A is a constant whose dimension is *reciprocal degree*. When pressure was held constant, in a similar manner volume varied with temperature,

$$V = V_0 (1 + At) \tag{2.13b}$$

where V is the volume at temperature t Celsius, and V_0 is the volume at 0°C. The constant A is the volume expansion coefficient for an ideal gas and has the value

$$A = \frac{1}{273.15} = 3.661 \times 10^{-3} \text{ K}^{-1}$$

The foregoing relationships are forms of Charles' law.

Since both Eqs. (2.13a) and (2.13b) are applicable, we get a common equation

$$PV = P_0 V_0 (1 + At)$$

This can be written

$$PV = P_0 V_0 A \left(\frac{1}{A} + t \right) \tag{2.14}$$

to give an insight into the process of expansion. When

$$t = -\frac{1}{A} = -273.15$$

the parenthetical term in Eq. (2.14) is equal to zero, and likewise P and V are zero. Because this value of t is the temperature at which a gas exerts no pressure, it is a basis for an absolute scale of temperature. Thus we have

$$T = \left(\frac{1}{A} + t\right) = 273.15 + t \text{ (in °C)}$$

where T is the temperature on the absolute scale, called the *Kelvin scale* in honor of the physicist. This scale is shown in Figure 2.1 for comparison with the other two.

In view of the foregoing we can write Eq. (2.14) as

$$PV = P_0 V_0 AT \qquad (2.15)$$

Now we let P_0 refer to standard pressure, V_0 refer to a molar volume V_A (a definite quantity of gas regardless of the species), and recall that A is the coefficient of expansion. The group of terms $P_0 V_0 A$ has a fixed value, and we represent it by a constant R_*,

$$R_* = P_0 V_0 A$$

A numerical value of R_* is obtained by substituting

$$P_0 = 1.0133 \times 10^5 \text{ N m}^{-2}$$
$$V_0 = V_A = 2.24 \times 10^{-2} \text{ m}^3$$
$$A = 3.661 \times 10^{-3}$$

This gives

$$R_* = 8.3143 \text{ J mole}^{-1} \text{ K}^{-1}$$
$$= 1.988 \text{ cal mole}^{-1} \text{ K}^{-1} \qquad (2.16)$$

Equation (2.15) then becomes

$$PV = R_* T \qquad (2.17)$$

which is the *equation of state* having wide applicability in all processes involving gases. The term R_* is the *universal gas constant* for a molar mass of any species and has the dimensions ML^2T^{-2}. Reading Eq. (2.17) with Eq. (2.16) in mind, it is clear that the quantity PV represents the innate energy of a molar mass at temperature T, and that R_* functions as a constant of proportionality.

The gas constant can be defined in terms of unit mass instead of molar mass. In such cases the constant has a different value for each species, depending on its molecular weight. For dry air, division of the value (2.16) by the assigned molecular weight 28.964 (number of grams in 1 mole of air), and conversion to a 1 kg mass, gives a *specific gas constant,*

$$R_{air} = 2.8706 \times 10^2 \text{ J kg}^{-1} \text{ K}^{-1}$$
$$= 6.8637 \times 10^4 \text{ cal kg}^{-1} \text{ K}^{-1} \tag{2.18a}$$

For water vapor whose molecular weight is 18, we have

$$R_v = 4.6191 \times 10^2 \text{ J kg}^{-1} \text{ K}^{-1}$$
$$= 1.1044 \times 10^5 \text{ kcal kg}^{-1} \text{ K}^{-1} \tag{2.18b}$$

When thus defined for unit mass the asterisk is not used with the symbol R. In the equation of state (2.17) the symbol V then represents the volume of unit mass of the gas, that is, the specific volume defined by Eq. (2.2). From this, the equation of state for unit mass can be written

$$P = \rho RT \tag{2.19}$$

which is useful in dealing with the free atmosphere where the gases are not confined to identifiable volumes.

2.2.2. Thermodynamic States and Processes

Several types of equilibrium states or conditions of a system are important in thermodynamics.

Thermal equilibrium means that the temperature is the same at all points within the system. If initial differences existed, heat has flowed from the hotter to the colder points, in accordance with the second law discussed in the following section, to produce a balance.

Mechanical equilibrium exists when internal forces or stresses have come into balance so that there is no detectable motion, on a macroscopic scale, of any part of the system. This is especially significant for a fluid in which all parts are free to move and reach equilibrium in response to applied forces. A large-scale example is a vertical column of air during a calm morning before sensible convection has started. Although the hydrostatic pressure in the column decreases as altitude increases, each layer is in mechanical equilibrium. The gravitational force (weight) on each layer is just countered by an equal upward force because of the pressure difference between its lower and upper surfaces. With respect to a fluid, mechanical equilibrium presupposes thermal equilibrium.

Chemical equilibrium means that, whatever the initial mixture of components, all possible chemical reactions have occurred. No further exchanges of reaction heat taken in (endothermic) or released (exothermic) take place. Thus chemical equilibrium is a necessary condition for thermal equilibrium.

When the above three forms exist, the system is considered to be in thermodynamic equilibrium. In discussing thermodynamic laws it will be assumed that the systems used as examples are in either thermodynamic equilibrium or (most helpfully) depart therefrom by only infinitesimal steps. Unless specified otherwise, the term *state* will imply one of equilibrium. The blackbody radiator discussed in Section 1.5.1 is a good example of a system in thermodynamic equilibrium.

To these traditional types of equilibrium we add two more because of their importance to absorption–emission processes. *Radiative equilibrium* means that the system—for example, a parcel of atmospheric air—is emitting as much energy, over all wavelengths, as it is receiving, over all wavelengths. This necessary consequence of Kirchhoff's law, Eq. (1.36), implies that the system temperature has risen or fallen to that value required by Planck's law, multiplied by the emissivity ϵ, to bring the two processes to equality from the total radiant energy standpoint.

The second type is *local thermodynamic equilibrium* (LTE), noted in Section 1.4.2 as the means whereby local energy excesses due to absorption, and local deficiencies due to emission, are spread by collisions over the immediate molecular population. This is the requirement for a single kinetic temperature of the gas: that all of the molecular internal energies be represented by the kinetic energy of translation, which is the quantity actually sensed by a thermometer. LTE is discussed further in Section 2.4.4.

Thus far we have described several kinds of equilibrium states, any of which can be considered stationary, so that thermodynamics might be called *thermostatics*. It is true that plenty of small-scale action is provided by kinetic theory, although this theory is not strictly a part of formal thermodynamics. The real escape from a completely static situation lies in the earlier statement that the system may depart from equilibrium by infinitesimal steps. These permit a process to take place that should be regarded as a succession of equilibrium states, each only slightly different from the preceding one.

Changes of state may take place by means of either *reversible* or *irreversible* processes, that is, along either two-way paths or one-way paths. At any point on a two-way path the direction can be reversed by a very small change in a variable such as temperature or pressure. As a simple example, visualize a closed container of gas fitted with a freely moving piston that can exert pressure on the gas. Now let an external force be applied to the piston. Equilibrium exists when the pressure exerted by the piston equals the internal pressure of the gas. The gas can now be expanded or compressed by reducing or increasing the piston pressure. It is shown in textbooks (see Fermi, 1937 or Wäser, 1966) that any closed system remains in equilibrium, and the process is reversible, if the change in the variable—in this case, the piston pressure—is made by infinitesimal steps.

Considering next an irreversible process, an easy example is seen in the sliding of one block against another. Because of friction some of the work is converted to heat, manifested in increased molecular motions of the block material and adjacent air. But we cannot reverse the process and convert this heat back into mechanical motions of the blocks. We have traveled a one-way path. An example of irreversibility closer to our subject is the expansion of a gas from a pressurized vessel to a connecting, evacuated vessel. After the pressure has equalized, the gas temperature and energy are unchanged from the initial values because no work has been performed, assuming that the gas is ideal. However, if left alone the gas will not reassemble itself into either vessel completely, with all the molecules in one and no molecules in the other. Here the system has definitely followed a one-way path. But the system disorder is greater after the expansion because the molecules, still with the same total energy as originally, now have a greater volume in which to move around. Our knowledge of their positions and momenta has become less exact.

2.2.3. Laws of Thermodynamics

The thermodynamic laws, actually postulates, have wide application and are often stated in various forms that emphasize particular aspects of the overall subject. The statements given here are intended to relate to our subject as directly as possible.

The *zeroth law* does not have a universal acceptance by this name. It was explicitly stated after the first law had already been assigned to the conservation of energy. The zeroth law states that two bodies that are separately in equilibrium with a third body are in thermal equilibrium with each other when brought together. This law represents a verifiable fact of experience and gives operational meaning to the concept of temperature and its measurement by a thermometer. Thus if two bodies produce identical readings on a thermometer, and for this purpose it need not be calibrated, then they are in thermal equilibrium with each other. This is an important role of a thermometer, that it act as a third or comparison body.

The *first law* states that the energy of any system remains constant when the system is isolated from all sources and sinks of energy. The system energy can be changed only when heat is transferred into or out of the system or when work is performed on the system or by it. The first law has been called the statement of the conservation of energy. Because the energy of an isolated system remains constant, some persons have concluded that the energy of the universe is constant. However, such a conclusion is based on cosmological assumptions not directly verifiable. Within the scope of the first law, "Energy can be neither created nor destroyed." Differently stated, the variation in energy of a system during any process or transformation is equal to the energy that the system receives from, or imparts to, its environment. The first law thus prohibits a perpetual motion machine that could perform work without a compensating change in an energy store somewhere else. Such an imaginary device is called a perpetual motion machine of the first kind because it would violate the first law.

The *second law* employs a somewhat elusive property called *entropy* and states that the entropy of an isolated system either remains constant or increases. It can never decrease. Qualitatively, entropy is considered to be a measure of the disorder or randomness of the system. A formal definition in terms of thermal quantities is given in Section 2.2.5, and additional insights from the viewpoint of statistical mechanics can be gained from Section 3.2.5. The disorder of the system lies in the many ways in which the molecules can be spread through volume–momentum space and through quantum number space. For example, the entropy of a liquid is greater than that of a solid, where the atoms occupy fixed points, while the entropy of a gas is still greater because the molecules move randomly in all directions. When a closed system is in equilibrium, the entropy has attained its maximum value for the given conditions. The prohibition that entropy never decrease also prohibits a perpetual motion machine that would produce work when the system and surroundings are at a uniform constant temperature. Always a transfer of heat is required if work is to be done, and the transfer must always take place from a higher to a lower temperature. Stated differently, no machine can be made that will solely transfer heat from a body of lower temperature to one of higher temperature without the expenditure of work. Any such imaginary device is called a perpetual motion machine of the second kind because it would violate the second law.

The *third law,* also called the *Nernst heat theorem,* states that the entropy of any well-ordered crystalline solid approaches zero as the absolute zero of temperature is approached. This would correspond to a state of complete order, meaning that any entropy differences between the possible configurations of a system have disappeared. We may consider that the configurations approach each other as the temperature is reduced, finally coalescing at the low-temperature limit. As pointed out by Wäser (1966), a perfectly ordered crystal at 0 K has only one microscopic state, because each atom is in a well-defined equilibrium position and is at its *zero-point* level of energy. Anticipating the Bose–Einstein statistics of Section 3.3.1, we may say that the atoms of the crystal, being identical, are indistinguishable. Hence any conceptual interchanges of the atoms cannot be recognized and do not produce additional microscopic states. We note in passing that, although temperatures near absolute zero have been reached in experiments, the absolute zero itself is not attainable (see Sears and Salinger, 1975).

2.2.4. Heat Capacity and Specific Heat

The first law of thermodynamics provides a good context for considering the quantities *heat capacity* and *specific heat capacity,* usually called *specific heat* as in this book. The heat capacity of any substance having any amount of mass is the amount of heat (energy) required to raise its temperature by 1°C. In SI parlance the specific heat of a substance is the amount of heat required to raise the temperature of unit mass of the substance by 1 K. Hence specific heat is the heat capacity per unit mass. Many writers, including Huntley (1958) in his treatise on dimensional analysis, define specific heat as the ratio of the heat capacity of a substance to the heat capacity of water. Many chemistry textbooks observe this traditional practice. Because the numerical values of the quantities have been chosen so that the heat capacity of water is 1 cal g^{-1} °C^{-1}, that is, *unity,* the specific heat of a substance has the same value by either definition. However, we should recognize that this alternate definition, which runs parallel to the definition of specific gravity, is conceptually different from the definition in SI terms.

For our purpose here we rephrase the first law of thermodynamics to read: *Any change in the energy of a system during a process is equal to the difference between the energy added to the system and the energy taken from it, for example, by doing work.* Now let heat in amount Q be added to unit mass of a gas while holding the volume constant. The system goes from P_1V_1 to P_2V_2 so that $\Delta P = P_2 - P_1$ and $\Delta V = 0$. All the heat added goes into molecular translational and internal energies since no external work is done. The system energy increases by $\Delta \mathscr{E}$ and the temperature by ΔT. The specific heat C_V of the gas, per unit mass, at constant volume is expressed by

$$Q = \Delta \mathscr{E} = C_V \, \Delta T$$

If the unit mass of gas is held at constant pressure while heat in amount Q is added, the system goes from P_1V_1 to P_1V_2 so that $\Delta P = 0$ and $\Delta V = V_2 - V_1$. A portion of the heat added goes into the work of expansion, represented by the product $P \, \Delta V$. The remainder of the heat added goes into molecular translational and internal energies, so that the system energy increases by $\Delta \mathscr{E}$ and the temperature by ΔT. The specific heat C_P of the gas, per unit mass, at constant pressure is expressed by

$$Q = \Delta \mathscr{E} + P \, \Delta V = C_P \, \Delta T$$

Because of the work term $P \, \Delta V$ we expect that C_P has larger values than does C_V.

From the discussion of the specific gas constant R in Section 2.2.1 we surmise that the specific heats described above have much in common with the specific gas constant R. Each of the quantities R, C_V, and C_P represents an amount of energy per unit degree, either Celsius or Kelvin, and per unit mass. For diatomic molecules at environmental temperatures we have the relationships

$$C_V = \frac{\Delta \mathscr{E}}{\Delta T} = 3\frac{R}{2} + 2\frac{R}{2} = \frac{5}{2}R \qquad (2.20)$$

where $R/2$ is attributed to each of the three degrees of translational freedom and $R/2$ to each of the two degrees of rotational freedom. For monatomic molecules (point masses) having no rotational freedom, the relationship is $C_V = 3R/2$. Looking next at C_P we have

$$C_P = \frac{\Delta \mathscr{E}}{\Delta T} + \frac{P \, \Delta V}{\Delta T} = 3\frac{R}{2} + 2\frac{R}{2} + R = \frac{7}{2}R \qquad (2.21)$$

This is the same as Eq. (2.20) except for the additional term R, which accounts for the work $P \, \Delta V$ done in expansion. For polyatomic molecules having three or more degrees of freedom, additional terms of $R/2$ enter the definition. At higher temperatures than those implied in the foregoing the vibrational degrees of freedom become active and increase the values of C_V and C_P still further. These additional mechanisms of energy storage are reviewed in Section 4.2.2.

Two sets of units for expressing the specific heats are in common use. In traditional thermodynamics the unit of heat is the gram-calorie, defined as the amount of heat required to raise the temperature of 1 gram of water by 1°C, over the range 14.5–15.5°C. In the cgs system of units the values of C_V and C_P are given in units of cal g^{-1} °C^{-1}. In the SI system the values are given in J kg^{-1} K^{-1}, which are also those of R as defined by Eq. (2.18a).

Table 2.2 lists the specific gas constant, the specific heats, and the dimensionless ratio C_p/C_V for the principal atmospheric gases. It is seen that C_p and C_v differ by approximately the value of R, which we expect from Eqs. (2.20) and (2.21). The ratio C_p/C_v is much used in studying gas properties, and an interesting application lies in the equation for the speed of sound in a gas,

$$v = \left(\frac{C_p}{C_v} \frac{P}{\rho}\right)^{1/2} \tag{2.22}$$

Substitutions from Eqs. (2.1) and (2.10a) and Table 2.2 yield a speed,

$$v = 332 \text{ m sec}^{-1} \tag{2.22a}$$

which agrees well with measured values. Since accurate measurements of this speed are readily made, they provide through Eq. (2.22) a check on the values of specific heats, which by their nature are difficult to measure accurately.

TABLE 2.2. SPECIFIC HEAT CAPACITIES OF ATMOSPHERIC GASES[a]

Constituent	In units of 10^2 J kg^{-1} K^{-1}			
	R	C_p	C_v	C_p/C_v
Dry air	2.871	10.04	7.170	1.40
Nitrogen	2.967	10.37	7.40	1.401
Oxygen	2.598	9.09	6.49	1.40
Argon	2.081	5.20	3.12	1.667
Carbon dioxide	1.889	8.20	6.30	1.30
Water vapor	4.615	18.47	13.86	1.333
Ozone	1.732	7.70	5.97	1.29
Helium	20.76	53.7	32.9	1.63
Hydrogen	41.24	142.5	101.3	1.407

[a]After Fleagle and Businger (1980).

2.2.5. Entropy and Phase Space

The idea of entropy was originated by Clausius in 1854 during the form-ative years of thermodynamics, which developed from early studies of the steam engine and efforts to improve its efficiency. Consider the operation of a heat engine that draws an amount Q_1 of heat from a source and in its operation transfers or discards an amount Q_2 to its environment (the cooler). The amount Q_2 is, to paraphrase Lord Kelvin, "irretrievably lost to man; having descended to the level of the environment, it is not annihi-lated but is completely wasted and can never again be reconverted to mechanical work." The amount the engine turns into useful work is $\Delta Q = Q_1 - Q_2$, and therefore the efficiency of the process over one or many cycles is

$$\frac{Q_1 - Q_2}{Q_1}$$

Associated with Q_1 and Q_2 are the temperatures T_1 of the source and T_2 of the cooler. It is shown in texts (for example, Mandl, 1980 and Sears and Salinger, 1975) that

$$\frac{T_1 - T_2}{T_1} = \frac{Q_1 - Q_2}{Q_1} \tag{2.23}$$

Thinking of the efficiency of operation as being defined by either side of this expression, it is clear that the efficiency will always be less than unity unless $T_2 = 0$ K. That is, unless the cooler, for example, the condenser in the case of a steam engine, is at absolute zero the efficiency must be less than 100%.

For the Carnot cycle of a heat engine we have

$$\frac{Q_1}{T_1} = \frac{Q_2}{T_2} \tag{2.24}$$

In words, the heat flowing in during the cycle, divided by the absolute temperature at which it flows in, is equal to the heat flowing out, divided by the absolute temperature at which it flows out. This ratio of flowing heat to temperature remains unchanged from the initial value at the end of a perfect cycle. Clausius recognized this ratio as a distinct thermal quantity which he called *entropy,* almost universally denoted by S. We recall that a thermodynamic process takes the system from one equilibrium state to another by way of infinitesimal steps, and that such steps are a requirement for the process to be reversible. Entropy then is formally defined for a reversible process by

$$dS = \frac{dQ}{dT} \qquad (2.25)$$

which means that, at the best, the entropy remains constant. For an irreversible process the entropy is defined by

$$dS > \frac{dQ}{dT} \qquad (2.26)$$

which means that the entropy increases as the process goes on. An isolated system, after the last addition or removal of heat, will always follow Eq. (2.26) in the direction of increasing entropy and reach an equilibrium state.

A classic example of an irreversible process is the expansion of a gas into an evacuated space, noted in Section 2.2.2. Following such expansion the right-hand side of Eq. (2.26) has not changed, because neither the energy nor the temperature of the gas has changed, but dS has increased because the occupancy of the larger volume by the same number of molecules has decreased the energy organization. Stated differently, the *nonavailable* energy has increased, as measured by the increase of entropy, and the molecular disorder also is greater. This last-named factor is important at this point. An equivalence between entropy and disorder is provided by statistical mechanics, as summarized in Section 3.2.4.

Here we get an advance, if qualitative, view of the equivalence from the concept of phase space, which is a domain available to molecules. In familiar, three-dimensional space the position of a molecule at a particular instant is specified by its xyz coordinates. In analogous manner the total momentum p (from which derives the energy) of the molecule at the same particular instant is specified by its three components p_x, p_y, p_z. Thus there are six defining quantities

$$x, \ y, \ z, \ p_x, \ p_y, \ p_z$$

for each molecule in a *phase space*, which we visualize as a six-dimensional continuum. A molecule at a particular point in phase space has a definite position and momentum. Therefore, the state of the system corresponds to a certain distribution of the molecules over the points in phase space.

The points in phase space, however, cannot be ideal mathematical points because of the uncertainty principle, Eq. (1.24). Rather, we must think in terms of very small six-dimensional cells whose sides are the differentials of the six quantities. The volume ΔV of each cell is given by

$$\Delta V = \Delta x \, \Delta y \, \Delta z \, \Delta p_x \, \Delta p_y \, \Delta p_z \qquad (2.27)$$

As we reduce the size of the differentials, the volume shrinks until the cell approaches the limit of a point. But from Eq. (1.24) we have

$$\Delta x \, \Delta p_x \geq h$$

$$\Delta y \, \Delta p_y \geq h$$

$$\Delta z \, \Delta p_z \geq h$$

where the factor 2π has been dropped because h is already very small.

Hence the phase cell volume defined by Eq. (2.27) is subject to the restriction

$$\Delta V \geq h^3 \qquad (2.28)$$

which effectively quantizes the spatial and momentum domain. Where, classically, position and momentum have continuous ranges to which the idea of infinitesimal increments applies, this is not true of phase space. The cell has finite size and at any instant each molecule is located somewhere within one of these cells. The number of cells evidently varies with the given volume of three-dimensional space and with the given momentum range. For these reasons the system disorder increases when a gas moves into a larger volume of space while retaining its initial energy and temperature. The larger number of phase cells simply means that there are more ways in which the molecules can be arranged among the cells. The "more ways" translates into more disorder, and, consequently, greater entropy. These ideas are carried further into the concepts of quantum states and their densities in Section 3.2, where a quantitative definition of entropy is given in terms of the number of possible molecular arrangements.

2.3. KINETIC THEORY OF GASES

The gas laws and principles of thermodynamics do not of themselves disclose the essential nature of a gas. That nature is revealed with detail and clarity by the *kinetic theory of gases,* developed over a span of two centuries by many workers. A beginning of the ideas may be sensed in the lively vision of Lucretius, as in the opening of this chapter. Quantitatively, the theory began with Bernoulli in 1738 and was carried forward in the 19th century by Dalton, Joule, Clausius, Maxwell, and Boltzmann. In 1910 Perrin demonstrated the reality of molecular motions by measuring the Brownian movements of colloidal particles suspended in a liquid. Still called a theory, this subject is based on an enormous amount of fact and is closely joined to thermodynamics and statistical mechanics.

In this and the following section we present those aspects of kinetic theory that are essential to an understanding of absorption–emission. Excellent treatments of the theory and its relation to thermodynamics will be found in Christy and Pytte (1965), Jancovici (1973), Loeb (1934) in detail, Rosenberg (1977), and Sommerfeld (1964). Advanced treatments which carry the reader into statistical mechanics are given by Gopal (1974), Incropera (1974), Jeans (1959), Kennard (1938), and Vincenti and Kruger (1965). Inquiring readers will gain insights from Kittel (1958, 1969) and Kittel and Kroemer (1980). Tutorially the treatment by Sears and Salinger (1975) is not surpassed.

2.3.1. Simplifying Assumptions, Elastic Collisions

In elementary kinetic theory it is usual to make simplifying assumptions about molecules and the nature of collisions. Generally, these assumptions lead to the concept of an ideal or perfect gas; the laws of Boyle and Charles and the equation of state apply without reservation to this ideal gas. Actually the atmospheric gases at STP conditions do not strictly meet all of the ideal gas requirements, primarily because of the large value of molecular number density at lower altitudes. As altitude increases, however, the number density exponentially decreases, and the ideal condition is realized at upper altitudes. For most purposes in atmospheric optics we can assume that atmospheric air behaves as an ideal gas, thus permitting a treatment that discloses some essential physics with minimal mathematics.

The assumptions are readily understood, and we recognize that some of them have been employed in preceding sections:

1. A gas consists of a great number of spherical molecules which move at high speeds in straight lines between collisions. Since they are material bodies, Newton's laws of motion apply to their movements and collisional interactions.

2. Their energies are regarded as entirely kinetic and nonquantized, ignoring at this point their internal quantized energies.

3. By ignoring internal energies, we thereby regard the molecules as perfectly elastic. With the sphericity assumption of (1), they thus are analogous to tiny billiard balls, although they must not have spin.

4. The molecules exert no forces on each other except during collisions when they elastically rebound. Momentum is conserved in the collisions, and the total kinetic energy of two or more colliding molecules remains unchanged by the collision.

5. The molecules are very small compared to their average separation distance. Stated differently, the total volume occupied by the molecules is not significant compared to the total volume of the gas. This means that the total volume of a gas, except at very high density, is mostly empty space. For example, 1 cm^3 of water at 100°C forms 1700 cm^3 of steam at standard pressure, so that the water molecules occupy only about 6×10^{-4} of the total volume of the vapor.

Two types of collisions are distinguished: *elastic* and *inelastic*. Here we emphasize the elastic type. The collision is elastic when it is considered to involve only translational kinetic energy, as when two nonspinning billiard balls collide and then move apart without spinning. Collision angles range from 0° (head-on) to 180° (grazing). Maxwell (1860) remarked about elastic collisions: "Instead of saying that the particles are hard, spherical, and elastic we may if we please say that the particles are centers of force, of which the action is insensible except at a certain small distance, when it suddenly appears as a repulsive force of very great intensity. It is evident that either assumption will lead to the same results." Elastic collisions are the means whereby translational kinetic energy is freely exchanged among all the molecules. Inelastic collisions, which involve exchanges between translational and internal energies, are discussed in Section 2.4.4.

2.3.2. The Molecular Quantities

The quantities listed here indicate the spatial and time scales of the molecular world. All numerical values refer to STP conditions where applicable. An air molecule has an *effective diameter*

$$d \approx 0.37 \text{ nm} = 3.7 \times 10^{-8} \text{ cm} \qquad (2.29)$$

This value is obtained from kinetic theory, but the concept of molecular diameter has several interpretations, as described later. Because Loschmidt's number N_L gives the number density per cubic centimeters, the average volume of space available to a molecule is $1/N_L$ cm^3. Hence the average spacing between molecular centers is approximated by

$$\bar{s} = \left(\frac{1}{N_L}\right)^{1/3} \approx 3.3 \text{ nm} = 3.3 \times 10^{-7} \text{ cm} \qquad (2.30)$$

or about nine times the molecular diameter. The *mean free path* $\bar{l}$, which is the average distance that a molecule travels between collisions, is given by

$$\bar{l} = \frac{1}{\sqrt{2}\,\pi d^2 N_L} \approx 60 \text{ nm} = 6 \times 10^{-6} \text{ cm} \qquad (2.31)$$

On the average only $\exp(-1)$ of the molecules travel this distance without suffering collisions. In contrast to the STP condition, at sea level for example, at an altitude of 125 km high in the ionosphere the value of number density N is only 10^{-8} the value of N_L, and the resulting mean free path is about 6 m. From Eqs. (2.10) and (2.19) we can write Eq. (2.31) as

$$\bar{l} = \frac{m}{\sqrt{2}\,\pi d^2 \rho} = \frac{mRT}{\sqrt{2}\,\pi d^2 P} \qquad (2.32)$$

to show that the mean free path varies directly as molecular mass and inversely as gas density, or directly as temperature and inversely as pressure. Thus from Eqs. (2.29) through (2.31) the molecular diameter, average spacing, and mean free path are in the ratios

$$d:\bar{s}:\bar{l} = 1:9:160$$

Because the effective range of intermolecular forces is of the order of molecular diameter, these ratios indicate that molecules interact only during collisions.

The average frequency of collisions per molecule, called the *mean collision rate* ϕ, is

$$\phi = \frac{\bar{v}}{\bar{l}} \approx 8 \times 10^9 \text{ sec}^{-1} \qquad (2.33)$$

The term $\bar{v}$ is the average or mean speed of the molecules as derived in Section 2.4.2. The mean free time τ between collisions, important in the collisional broadening of spectral lines, is defined by

$$\tau = \frac{1}{\phi} = \frac{\bar{l}}{\bar{v}} \approx 10^{-10} \text{ sec} \qquad (2.34)$$

It is interesting to apply the concept of mean free time (and mean free path) to a multitude of molecules and see the distribution of statistical values about the mean value. Consider that N_0 molecules (a reference number) are involved. Out of the original number N_0, the number N which have not collided after traveling for time t is given by

$$N = N_0 \tau^{-1} \exp\left(-\frac{t}{\tau}\right) \qquad (2.35)$$

Hence the relative number which have not suffered collisions during time t is

$$\frac{N}{N_0} = \tau^{-1} \exp\left(-\frac{t}{\tau}\right)$$

which is plotted in Figure 2.4 in terms of the parameter t/τ. For example, at the end of the period $t = \tau$, only 36.8% of the molecules have not undergone collisions. Equation (2.35) is called the *survival equation,* and a similar relationship holds for the mean free path.

2.3.3. Translational Energy, Boltzmann's Constant

The translational kinetic energy of a molecule at any instant is a function only of its mass and speed. This energy is not quantized and, as for a macroscopic body, is defined by

$$\mathscr{E}_{tr} = \tfrac{1}{2}mv^2 \qquad (2.36)$$

where v is the instantaneous speed. Molecular speeds are isotropic because

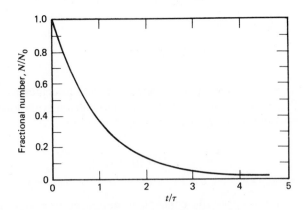

Figure 2.4.   Fractional number of molecules traveling from time t in a gas at STP conditions without suffering collisions, as a function of the ratio t/τ.

each molecule has three degrees of translational freedom along coordinate axes that can have any orientation and because each molecule moves randomly. This is a form of the principle of equipartition of energy, that the average energy is equally divided along any three orthogonal axes.

When a gas is confined, the macroscopic pressure exerted against the container wall is an average effect from the reversals of momenta as the gas molecules elastically rebound from the molecules constituting the container wall. This pressure is defined in terms of purely molecular quantities by

$$P = \tfrac{1}{3}Nm\overline{v^2} \tag{2.37}$$

where N is the number density, m is the molecular mass, and $\overline{v^2}$ is the mean-squared speed. The factor $\tfrac{1}{3}$ accounts for the equipartitioning of energy along the three axes.

The root-mean-square (rms) molecular speed in terms of gas bulk properties is found from Eq. (2.37). Substituting for the product mN its equivalent quantity ρ from Eq. (2.10), we get

$$P = \tfrac{1}{3}\rho\overline{v^2} \tag{2.38}$$

Thus the rms speed is

$$v_{\text{rms}} = (\overline{v^2})^{1/2} = \left(\frac{3P}{\rho}\right)^{1/2} \tag{2.39}$$

Substituting therein the previously stated values for STP air:

$$P = 1.0133 \times 10^5 \text{ N m}^{-2}$$
$$\rho = 1.29 \text{ kg m}^{-3}$$

we find

$$v_{\text{rms}} = 485 \text{ m sec}^{-1} \tag{2.40}$$

This is somewhat comparable to the speed of sound, which is 332 m sec^{-1} according to Eq. (2.22a). This result is not surprising because directed molecular motions are the means for propagating acoustic waves. The mean kinetic energy of a single air molecule is found by substituting into Eq. (2.36) the values of m from Eq. (2.8a) and v_{rms} from Eq. (2.40). The result is

$$\mathscr{E}_{\text{tr}} = 5.657 \times 10^{-21} \text{ J} \tag{2.41}$$

We next examine the role of the molecule in the equation of state. Rewriting Eq. (2.38) in the light of Eq. (2.10),

$$PV_A = \tfrac{1}{3}M_A\overline{v^2} = \tfrac{1}{3}N_A m\overline{v^2} \tag{2.42}$$

and combining this with Eq. (2.17), we obtain

$$3PV_A = M_A\overline{v^2} = 3R\cdot T \tag{2.43}$$

The total translational kinetic energy inherent in the molar mass M_A resides in the kinetic energies of N_A molecules, each having mass m and speed v_{rms} defined by Eq. (2.39). Thus the energy of the molar mass is expressed by

$$\tfrac{1}{2}M_A\overline{v^2} = \tfrac{1}{2}N_A m\overline{v^2} = \tfrac{3}{2}R\cdot T \tag{2.44}$$

The mean kinetic energy $\mathscr{E}_{tr}$ of a single molecule is found by dividing Eq. (2.44) by N_A. Using only the last two terms this gives

$$\overline{\mathscr{E}_{tr}} = \frac{1}{2}m\overline{v^2} = \frac{3}{2}\frac{R\cdot T}{N_A} = \frac{3}{2}\kappa T \tag{2.45}$$

where the first group of terms corresponds to Eq. (2.36). Thus the mean kinetic energy is a function of temperature only, and this is true for any molecular species. The rms speed is defined in terms of mass and temperature by

$$v_{rms} = (\overline{v^2})^{1/2} = \left(\frac{3\kappa T}{m}\right)^{1/2} \tag{2.46}$$

In Eq. (2.45) the quotient $R\cdot/N_A$ represents *Boltzmann's constant*, usually denoted by κ, which has the value

$$\kappa = 1.381 \times 10^{-23} \text{ J K}^{-1} \tag{2.47}$$

and dimensions ML^2T^{-2}. Because the specific gas constant R, defined for air by Eq. (2.18a), refers to unit mass we have

$$\kappa = Rm \tag{2.48}$$

A nontrivial use of this relation is its application to the last part of Eq. (2.32), which becomes

$$\bar{l} = \frac{\kappa T}{\sqrt{2}\,\pi d^2 P} \tag{2.49}$$

This expression is important in determining the collisional broadening of spectral lines as a function of T and P.

From Eq. (2.45) we see that Boltzmann's constant defines the mean amount of kinetic energy that a molecule can claim as a function of temperature. When a gas is in thermal equilibrium, molecules of different masses have different mean speeds to satisfy Eq. (2.46), but the total energy of the gas is equally divided among all the molecules, and these equal mean energies are functions only of temperature and Boltzmann's constant. From this we see that temperature is more than an abstraction and more than a sensory perception. Temperature, however defined and on whatever scale, is the thermodynamic quantity that is proportional to the kinetic energy of molecular translation.

2.4. DISTRIBUTIONS OF MOLECULAR POPULATIONS

Molecular velocities, speeds, and the resulting kinetic energies cover wide ranges of values. The total molecular population is distributed over these ranges according to functions developed by Maxwell from probability considerations and by Boltzmann from thermodynamic reasoning. In this section we review these three functions because of their relevance to the broadening of spectral lines, and because they provide an insight into molecular behavior. They give us an introductory view of the Boltzmann distribution of particles over energy states, a subject that is developed further in Chapter 3 with the aid of statistical mechanics. Here the treatments of the functions stress explanation rather than derivation; readers interested in derivations may consult among the texts cited at the beginning of Section 2.3.

2.4.1. Velocity Function

Velocity space for the molecules of a gas in thermal equilibrium is isotropic. Along any direction of observation the velocity components parallel to that direction produce Doppler shifts in the wavelengths of the energy absorbed and emitted by the molecules. These shifts are responsible for the Doppler broadening of spectral lines, hence our concern with the velocity function.

Consider now a unit volume of gas consisting of N identical molecules moving at random speeds in random directions as a result of collisions. At any instant each molecule has a total velocity (speed) that is specified by its three components along a set of orthogonal axes having any orientation. The number dN of molecules having velocity components in the intervals

$$v_x \text{ and } v_x + dv_x$$
$$v_y \text{ and } v_y + dv_y$$
$$v_z \text{ and } v_z + dv_z$$

is defined by

$$dN = N\left(\frac{m}{2\pi\kappa T}\right)^{3/2} \exp\left[-\frac{m}{2\kappa T}\left(v_x^2 + v_y^2 + v_z^2\right)\right] dv_x\, dv_y\, dv_z \quad (2.50)$$

where the group of terms forming the multiplier of N is, in effect, a probability distribution normalized to unity.

Fixing attention on one set of components, for example, v_x, we have from Eq. (2.50) that

$$dN = N\left(\frac{m}{2\pi\kappa T}\right)^{1/2} \exp\left(-\frac{m}{2\kappa T}v_x^2\right) dv_x \quad (2.51)$$

Here dN is the number of molecules having an x component of velocity between v_x and $v_x + dv_x$, and the ratio dN/N that can be formed represents the fractional population occupying this interval. If the gas is in thermal equilibrium, the rate at which molecules enter this interval by collisions is equal to the rate at which they leave it by collisions.

Equation (2.51) has the form of a Gaussian error distribution (normal probability) function, whose equation is

$$y = \frac{1}{\sqrt{\pi}\,\sigma} \exp\left(-\frac{x^2}{\sigma^2}\right) \quad (2.52)$$

where y = ordinate value at given value of x
      σ = standard deviation of the distribution, equal to the rms value of the deviations from the mean
      x = deviation of the variable from the mean
If we let

$$\sigma = \left(\frac{2\kappa T}{m}\right)^{1/2} \quad \text{dimensions: } LT^{-1}$$

$x = v_x$ dimensions: LT^{-1}

$y = \dfrac{dN}{N}\dfrac{1}{dv_x}$ dimensions: $L^{-1}T$

and substitute into Eq. (2.52) we get an expression identical to Eq. (2.51). We note that the parameters defining σ are Boltzmann's constant, gas temperature, and molecular mass. Assuming an air temperature of $0°C = 273.15$ K, we substitute for m and κ their values from Eqs. (2.8a) and (2.47) to get

$$\sigma = \left(\frac{2\kappa T}{m}\right)^{1/2} = 396 \text{ m sec}^{-1} \tag{2.53}$$

for the standard deviation of velocity at a given temperature.

Figure 2.5 shows a plot of Eq. (2.51) for air molecules at $0°C$, with dN/dv_x as the right-hand ordinate scale and v_x and v_x/σ as dual abscissa scales. The area under the curve represents the total probability whose value is unity. Arranging the figure for easy visualization of dN/dv_x, we let this area represent also the total number density N, assumed to be 10,000. Assuming $dv_x=1$ for convenience, the area of any differential strip having height dN and width dv_x is then dN. Thus the left-hand ordinate scale, which is the product of N and dN/dv_x, indicates the numerical value of dN for these

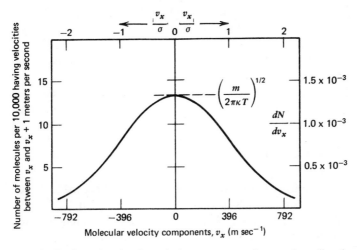

Figure 2.5. Distribution of molecular velocity components in any given direction for air molecules at $0°C$.

assumptions. The most probable value of v_x corresponds to the maximum of the curve, which occurs at $v_x=0$. For this value, Eq. (2.51) becomes

$$\frac{dN}{dv_x} = N\left(\frac{m}{2\pi\kappa T}\right)^{1/2}$$

The parenthetical term has the value 1.42×10^{-3} for air at 0°C. With $N = 10,000$ and $dv_x=1$ m sec^{-1}, we find that approximately 28 molecules have velocity components between $v_x=-1$ and $v_x=1$ m sec^{-1}.

The symmetry of the curve for positive and negative v_x argues that as many molecules are moving toward an observer as are moving away from him. In the event, this must be happening. It is characteristic of the Gaussian distribution that 68.26% of the values lie between the -1σ and the 1σ points. Thus out of the total 10,000 molecules, about 6826 have velocity components between $v_x=-396$ and $v_x=396$ m sec^{-1}. Likewise, 95.44% are between -2σ and 2σ, and 99.74% are between -3σ and 3σ. The effect of increasing the temperature is to increase the value of σ, which lowers and widens the distribution.

That the most probable value of v_x is zero may seem unrealistic, but this is only one of three orthogonal components of velocity. Those molecules having v_x near zero may have any values of v_y and v_z because the three probabilities are independent. It is rare that all three components are near zero at the same instant for a particular molecule. Figure 2.6 attempts to suggest this with a few molecules randomly distributed as dots in two-dimensional velocity space. The coordinates of each dot were selected from random numbers normally distributed around zero. Each dot is a combination of random v_x and v_y, and the magnitude of the total velocity v is equal to the length of a line from the origin to the dot. Although there are many dots having one or the other of the components near zero, there are relatively few with v itself near zero. Now in imagination extend this situation to three-dimensional velocity space by introducing the component v_z. It then can be visualized that still fewer molecules will have a total velocity near zero.

2.4.2. Speed Function

The distribution of molecular velocities without regard to direction is the speed distribution, which is found by extending Eq. (2.51) over three-dimensional velocity space. The absolute value of velocity in terms of its three components is defined by

$$v = (v_x^2 + v_y^2 + v_z^2)^{1/2}$$

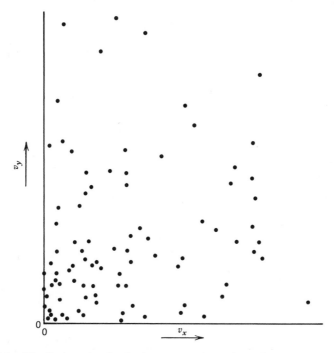

Figure 2.6. Distribution of molecules in one quandrant of two-dimensional velocity space.

The number dN of molecules having speeds between v and $v + dv$ is given by

$$dN = N\frac{4v^2}{\sqrt{\pi}} \left(\frac{m}{2\kappa T}\right)^{3/2} \exp\left(-\frac{m}{2\kappa T}v^2\right)dv \qquad (2.54)$$

The ratio dN/N defines the fraction of the total population having speeds in this interval. If the gas is in thermal equilibrium, the rate at which molecules enter this interval is equal to the rate at which they leave it. This function also is normalized to give a total probability of unity when integrated over all velocities. Figure 2.7 shows a plot of Eq. (2.54) for oxygen molecules at $0°C$ and $-200°C$. The ordinate unit is dN/dv, where dv is taken as a differential quantity equal to unity. The area of any strip is equal to the product of its height dN/dv and its width dv (unity). The areas under the two curves are equal and proportional to N, taken to be 10,000 as for the velocity function in Figure 2.5.

Three concepts of speed are noted in Figure 2.7. The most probable

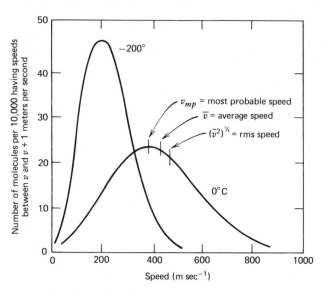

Figure 2.7. Distribution of the molecular populations over the speed range for oxygen at 0°C and −200°C. From Osgood et al., (1964) and McCartney (1976).

speed v_{mp} corresponds to the maximum of the curve and is found from the derivative of Eq. (2.54). This turns out to be

$$v_{mp} = \left(\frac{2\kappa T}{m}\right)^{1/2} \tag{2.55}$$

For air at 0°C, substitutions into Eq. (2.55) give

$$v_{mp} = 396 \text{ m sec}^{-1} \tag{2.56}$$

Interestingly this is the same value found for the 1σ value of v_x in Eq. (2.53). The function (2.54) is such that the probability of a speed greater than $4v_{mp}$, for example, is only about 10^{-6}.

The average or mean speed $\bar{v}$ is found from Eq. (2.54) by the method usually employed to obtain a weighted average of a variable having a defined distribution. To do this, we multiply each value of v by the number dN of molecules having that speed, integrate over the total range of speeds, and divide by the total population N. This procedure yields

$$\bar{v} = \left(\frac{8\kappa T}{\pi m}\right)^{1/2} = 1.13 v_{mp} \tag{2.57}$$

The expression for the rms speed is obtained from Eq. (2.54) as

$$v_{rms} = \left(\frac{3\kappa T}{m}\right)^{1/2} = 1.22 v_{mp} = 1.08 \bar{v} \qquad (2.58)$$

For air at 0°C = 273.15 K, substitutions from Eqs. (2.8a) and (2.47) into Eq. (2.58) give

$$v_{rms} = 485 \text{ m sec}^{-1} \qquad (2.59)$$

which agrees with the value in Eq. (2.40) obtained from the macroscopic properties. The effect of gas temperature on the shape of the distribution curve should be noted in Figure 2.7.

2.4.3. Energy Function, Boltzmann Factor

In dealing with absorption–emission it is important to know the manner in which the molecules are distributed over the total range of energies. Of particular interest is the clustering of energies about the mean value $3\kappa T/2$ given by Eq. (2.45). For this purpose the speed distribution, Eq. (2.54), is converted to an energy distribution by a change of variable. From Eq. (2.36) we get $dv = d\mathscr{E}/mv$ and then rewrite Eq. (2.36) as

$$v = \left(\frac{2\mathscr{E}}{m}\right)^{1/2}$$

Making these substitutions in Eq. (2.54) gives

$$\frac{dN}{d\mathscr{E}} = 2N\left[\frac{\mathscr{E}}{\pi(\kappa T)^3}\right]^{1/2} \exp\left(-\frac{\mathscr{E}}{\kappa T}\right) \qquad (2.60)$$

which is plotted in Figure 2.8. The interpretation of this curve is similar to that of the speed curve in Figure 2.7. The most probable value of energy corresponds to the curve maximum found from the derivative of Eq. (2.60), and is

$$\mathscr{E}_{mp} = \tfrac{1}{2}\kappa T$$

so that the quantity κT is a natural abscissa unit for Figure 2.8. The average energy of any one of the molecules was found by Sproull (1963) to be

$$\bar{\mathscr{E}} = \tfrac{3}{2}\kappa T$$

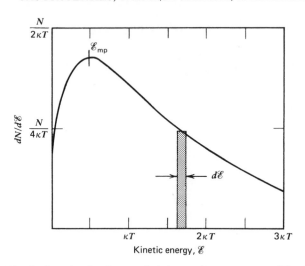

Figure 2.8. Distribution of molecular population over the energy range. The shaded area is proportional to the number of molecules having kinetic energies between ε and $\varepsilon + d\varepsilon$. After Sproull (1963).

This result, which is a weighted average because we are dealing with a distribution, agrees with Eq. (2.45).

At this point we note the forms of the distribution functions for energy and speed. Looking at Eq. (2.60), the population in the interval between $\mathscr{E}$ and $\mathscr{E} + d\mathscr{E}$ is increasingly controlled by the exponential term as $\mathscr{E}$ takes on larger values. This is evident in Figure 2.8 where we see the exponential overtaking the multiplying term $\mathscr{E}^{1/2}$ at the value $\mathscr{E}_{mp}$ and then surpassing it. An equivalent effect occurs for the speed function (2.54) due to the factor v^2, as may be seen in Figure 2.7. The argument of the exponential term in these functions is the ratio of a given value of energy to the statistical value set by κT. This is the essence of the Boltzmann distribution as employed in statistical mechanics: The probability that a molecule will be at a particular energy level $\mathscr{E}_1$ is a function of the term

$$\exp\left(-\frac{\mathscr{E}_1}{\kappa T}\right)$$

This is called the *Boltzmann factor* and is employed frequently in subsequent explanations.

Boltzmann's distribution has many applications in thermodynamics, physics, and chemistry. It is not restricted to molecular entities but embraces all bits of matter from electrons to colloidal and aerosol particles. The relevant energy may be that of molecular translation or the internal energy of a molecule or the potential energy of a molecule in the gravitational field. In considering two energy levels $\mathscr{E}_1$ and $\mathscr{E}_2$ that are accessible to a molecular population, with $\mathscr{E}_1 < \mathscr{E}_2$, the relative populations of the two levels conform to

$$\frac{N_2}{N_1} = \exp\left(-\frac{\mathscr{E}_2 - \mathscr{E}_1}{\kappa T}\right) \tag{2.61}$$

and so on for any two levels. This is Boltzmann's distribution in elementary form and leads to the general conclusion that, when matter has freedom of arrangement or of combination, the most probable resulting state is the one with the lowest total energy. This is the equilibrium state toward which a system moves in maximizing the entropy. On the microscopic scale this principle governs the populations of speed and energy intervals, as we have seen. On an equivalent scale it governs the populations of internal energy levels, as treated later.

On a macroscopic scale Boltzmann's principle governs the vertical distribution of the atmosphere, which exists only by maintaining itself in kinetic suspension against the force of gravity. We review this matter briefly because it is a good, observable example of the Boltzmann distribution. Assuming that the atmosphere from top to bottom is in an isothermal state and that it is in hydrostatic equilibrium, the pressure at any point is due to the downward force (weight) exerted by the overlying column of air. This relationship is expressed by the hydrostatic equation

$$-dP = \rho g \, dZ \tag{2.62}$$

where dP is the increment in pressure due to an increment dZ in altitude. Substituting for ρ its value from Eq. (2.19) we have

$$-dP = \frac{Pg}{RT} \, dZ \tag{2.63}$$

Integration between the limits P_0, where the altitude is zero, and P_z where the altitude is Z, yields

$$P_z = P_0 \exp\left(-\frac{gZ}{RT}\right) \tag{2.64}$$

Equation (2.64) is the *law of atmospheres,* originally derived by Laplace.

Substituting for R its value from Eq. (2.48) we get the law in terms of molecular quantities,

$$P_z = P_0 \exp\left(-\frac{gmZ}{\kappa T}\right) \tag{2.65}$$

The numerator of the exponential argument is now identified as the potential energy of a molecule due to its height above the earth's surface, while the denominator represents the average kinetic energy of that molecule. Making use of the relationships

$$\frac{P_z}{P_0} = \frac{\rho_z}{\rho_0} = \frac{N_z}{N_0}$$

obtained from Eqs. (2.10) and (2.19), we rewrite Eq. (2.65) as

$$P_z, \rho_z, \text{ or } N_z = (P_0, \rho_0, \text{ or } N_0) \exp\left(-\frac{gmZ}{\kappa T}\right) \tag{2.66}$$

to show that pressure, mass density, and number density have the same exponential dependence on altitude in an isothermal atmosphere.

In the exponential argument of Eq. (2.66) the group of terms $gm/\kappa T$, where T is the variable, acts as a scaling factor to control the rate of exponential decrease as Z increases. Thus temperature is the governing parameter. When $Z = \kappa T/mg$ the value of P_z, ρ_z, or N_z is just $\exp(-1) = 36.8\%$ of the value at $Z = 0$. Hence a *scale height H* for an isothermal atmosphere can be defined by

$$H = \frac{\kappa T}{mg} \tag{2.67}$$

Assuming a standard temperature $T = 273.15$ K ($= 0°$C), and substituting for κ, m, and g from Appendix A, we get for the scale height,

$$H = 8.00 \times 10^3 \text{ m} \tag{2.68}$$

or about 26,000 ft. Scale height, or *scale factor* generally, is a useful parameter of exponential distributions.

The isothermal assumption should be employed judiciously. Although several altitude regions, including the stratosphere, are approximately isothermal, the overall atmosphere is not. In the troposphere alone, for example, the temperature lapse rate defined by Eq. (2.3) has the relatively large value 0.65°C per 100 m. Nevertheless, the isothermal model is useful in studying the vertical distribution of atmospheric properties, as are two others called the *constant density* and the *polytropic* models, which are not isothermal. These three analytic models are described by McCartney (1976), and more fully by Tverskoi (1965). The widely used *standard model* is USSA (1976), which embodies extensive tabulations of annual mean values of atmospheric properties for all altitude regions, along with a detailed text. This supersedes the model USSA (1962), which is still valuable, as are the seasonal supplements USSAS (1966).

2.4.4. The Reservoir of Energy

Because of their translational motions, gas molecules act as reservoirs of kinetic energy, which is freely exchanged by the elastic collisions defined in Section 2.3.1. In this section we consider *inelastic collisions* by which translational and internal energies are exchanged. This is a basic phenomenon by which energy in radiant form is imparted to and taken from a mass of gas. These exchanges between the translational and internal forms—rotational, vibrational, and electronic—are responsible for local thermodynamic equilibrium (LTE), mentioned in Section 1.4.2 as a requirement for Kirchhoff's law Eq. (1.35) or Eq. (1.36) to apply. We now look at the mechanism of the collisional exchanges of energy.

There are two kinds of inelastic collisions. In the first kind, internal energy is gained at the expense of translational. This is collisional activation or excitation in which a molecule undergoes a nonradiative transition to higher-energy levels. If the molecule then undergoes a radiative transition downward, evidently the gas loses energy by this exchange and emission. Therefore its temperature decreases. In the second kind of inelastic collision, translational energy is gained at the expense of internal energy. This is collisional deactivation or deexcitation. If the internal energy thus converted to kinetic energy was gained from a radiative transition upward, the gas has gained energy by absorption. Therefore its temperature increases.

Collisional deactivation, known as *relaxation of internal energy,* involves mostly the rotational and vibrational forms, hence its importance in the infrared. The collisional rate given by Eq. (2.33) is about 8×10^9 sec^{-1} for STP conditions, equivalent to approximately 10^{-10} sec between collisions. Relaxation rate is equal to the product of collision rate and the probability of deactivation, while relaxation time is equal to the reciprocal of this product. Although the probabilities of deactivation are low, they are sufficiently compensated by high collision rates to yield short relaxation times. If these are less than the natural lifetimes of excited states, the internal energy gained from an absorptive transition is more likely to be thus converted to translational energy than to be emitted directly. These conversions add to the store of kinetic energy and in so doing they increase the gas temperature. This energy store is then available for collisional activation of any nearby molecule, which subsequently will undergo an emissive transition or still another collisional deactivation.

Relaxation data for molecules in the atmosphere are sparse, but some estimates are given by Goody (1954, 1964) and Curtis and Goody (1956). In the lower atmosphere the relaxation times are far shorter than the natural lifetimes of excited rotational and vibrational states. For example, certain vibrational states of H_2O and CO_2 molecules have relaxation times of 10^{-6}–10^{-5} sec, and rotational states have even shorter relaxation times. In contrast, the natural lifetimes of vibrational states are near 10^{-1} sec, while those of rotational states are somewhat longer. Where vibrational–rotational transitions can occur, however, the natural lifetime tends to be set by the shorter vibrational lifetime. Goody (1964) estimates that the ratio of relaxation time to natural lifetime may be of the order 10^{-10} sec at STP conditions for rotational levels, and 10^{-5}–10^{-4} sec for vibrational levels.

Hence the initially absorbed energy is distributed among the molecules by a series of collisional deactivations, purely kinetic energy transfers, and collisional activations before it is finally emitted. In this manner inelastic and elastic collisions act jointly to spread the received energy over the entire population, preventing an excess from piling up here and a deficiency from developing there. These processes give each momentum state and each quantum state its due statistical share of the total thermal energy, thus *maximizing the entropy* and creating LTE. From the numbers cited above, it appears that LTE prevails up to altitudes of 70 km or so. At higher altitudes the number density N becomes much smaller, as shown by Eq. (2.66). The resulting free path $\bar{l}$, defined by Eq. (2.31), then is sufficiently great that the collision rate φ, defined by Eq. (2.33), can no longer maintain LTE.

A very large number of quantum energy states are available to gas molecules, as discussed in the following chapter. The important consequence of LTE is that the mean populations of these states, as well as the momentum states noted in Section 2.2.5, are controlled by gas temperature through molecular collisions as just described. When LTE exists, a single *kinetic temperature* applies to all molecules of the gas, and this temperature represents all the forms of molecular energy. The gas then obeys Kirchhoff's law and exhibits blackbody characteristics when the optical thickness, defined by Eq. (1.31), is great and graybody characteristics when it is small. In the lower atmosphere collisional exchanges between translational and internal energies are limited principally to the rotational and vibrational forms.

Collisional activation and deactivation of electronic states occur infrequently because the values of electronic energy are so much greater than those of translation energy that the two forms do not easily interchange. Deactivation of these states apparently is accomplished by a series of cascaded vibrational transitions. Conditions are otherwise, however, in the ionosphere at altitudes between 80 and 500 km or so. Solar particles of high energy collisionally excite the electronic states of N_2 and O_2 molecules, in addition to producing the ionizations for which this altitude region is named. The natural lifetimes of these excited states, which are extremely unstable because of their high energies, are only about 10^{-8} sec, but the relaxation times at such altitudes are considerably greater. Emission of visible light then occurs directly from these states, indicating that this atmosphere is not in a condition of LTE. When the solar bombardment is intense and is directed into high latitudes by interactions with the geomagnetic field, spectacular auroral displays are created.

3

Quantized Energy States and Populations

The probability that we assign to an event is always relative to the knowledge (actual or presumed) that we possess. [Eddington (1935)]

A consequence of Heisenberg's uncertainty principle is that the behavior of individual bits of matter in very small domains cannot be predicted exactly. With a multitude of particles and an even greater number of domains, we must deal with the behavior probabilities of both the individual and the group. We saw this to be true for molecular velocities and translational energies which, although individually unpredictable, do conform to probability distributions in the aggregate of large numbers. From these probabilities come reliable statistical predictions of macroscopic (observable) behavior of a gas system as its physical parameters are altered.

A similar situation exists with the quantized states of internal energy and their chances of being occupied by a swarm of molecules. Here the methods of statistical mechanics provide a sure guide to finding the most probable distribution of the population. It will be seen that the occurrence probability of a particular arrangement is so high that it amounts to a virtual certainty. Thus, in line with Eddington's comment, our knowledge (statistical) has become very great through the power of statistical mechanics.

In this chapter we review the concepts of quantized energy states and the distributions of molecular populations over these states. Simple physical models are employed for illustration. The standing waves of classical physics, de Broglie matter waves, and the Schrœdinger wave functions are discussed. Next, the quantum states are defined, and the number within a given volume is enumerated. System microstates and macrostates are explained, and the relation between entropy and information is noted. Employing the methods of statistical mechanics, three population distributions —Bose–Einstein, Fermi–Dirac, and Boltzmann—are analyzed and applied to a small system. The chapter concludes with a review of the indices usually employed to define the spread of a molecular population over the accessible energy levels.

These are indeed extensive subjects to treat in one chapter, and we suggest further reading. Texts which have been found helpful are Beiser

(1967), Eisberg and Resnick (1974), Kittel and Kroemer (1980), and Mandl (1980). Sears and Salinger (1975) and Vincenti and Kruger (1965) are particularly valuable in dealing with the statistics of distributions and calculating their most probable values. Virtually all aspects of the material covered in this chapter are treated in detail by Sommerfeld (1964) and Tolman (1938), and a careful study of their works will be rewarding.

3.1. WAVES IN CLASSICAL AND QUANTUM MECHANICS

One of the great unifying ideas of physics is the wave concept, which embraces such seemingly diverse phenomena as sonic propagation, vibrations of an elastic cord, rippling surface of a lake, electromagnetic waves, and probability waves in quantum mechanics. A common characteristic of all types of waves is that a variation in some property of a medium is propagating. We may regard such a variation as a disturbance or anomaly. As mentioned in Appendix C with respect to electromagnetic energy, the disturbance cannot be confined to its point of origin. The disturbance must move, becoming a function of time at any point on the path, and a function of distance at any instant of time. Hence the wave exists in both time and space as it propagates. The wave may be either continuous, as a smoothly varying function of time and distance, or it may be discontinuous, as a wave packet or pulse.

In this section we review the properties of classical standing waves to provide a background for considering the types of waves that are central in quantum mechanics. The idea of a matter wave that accompanies a moving particle is then discussed. This leads to the Schroedinger equations, which are illustrated by "a particle trapped in a box," that is, a particle bound by external forces to a small volume of space. The common factors of waves in classical and quantum mechanics are emphasized.

3.1.1. Classical Standing Waves

We start by considering the vibrations of an elastic cord stretched between rigid supports, as illustrated in Figure 3.1. The vibrations take the form of standing or stationary waves whose wavelengths bear fixed relationships to the length and tension of the cord. In elementary theory it is shown that the standing wave is formed by the superposition of two component waves traveling in opposite directions. Each wave is reflected at an anchor point with a phase change of π rad, and the resulting interference

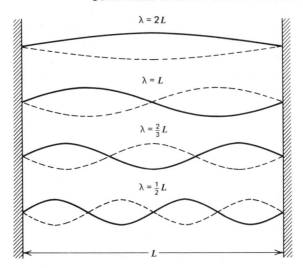

Figure 3.1. Vibrational modes of a stretched elastic cord. From Beiser (1967). *Concepts of Modern Physics.* © 1963 by McGraw-Hill, Inc. Used with the permission of McGraw-Hill Book Co.

between the two components produces a stationary wave whose equation is

$$y = 2A \, \cos(-kx)\sin(\omega t) \tag{3.1}$$

Here y is the instantaneous displacement of a point on the cord and A is the amplitude or maximum displacement. We see that y is a function of position x along the cord, according to the term $\cos(-kx)$, where k is the propagation constant defined by $k = 2\pi/\lambda$. Also y is a function of time t according to $\sin \omega t$, where ω is the angular frequency of the reference point associated with simple harmonic motion.

The wave of Eq. (3.1) obviously cannot travel but remains confined between the anchor points of the cord. The maximum value of y varies as $\cos(-kx)$ between the value of A, where $kx = 0$, π, 2π, ... , and zero where $kx = \pi/2$, $3\pi/2$, $5\pi/2$, ... The first group of positions marks the antinodes or points of maximum displacement, which are separated by the distance $\Delta x = \pi/k = \lambda/2$. The second group of positions marks the nodes or points of zero displacement, also separated by $\Delta x = \pi/k = \lambda/2$. The instantaneous value of y, at any position x, varies as $\sin \omega t$ between the maximum values permitted by $\cos(-kx)$.

The tension and mass of the cord, along with the physical constraint, produce a resonant vibrating system. We note that the velocity of either component wave is given by

$$v = \left(\frac{T}{m}\right)^{1/2}$$

where T is the tension of the cord and m is its mass per unit length. Although numerous modes of resonant vibration are possible, as suggested by the figure, the anchor points must always be nodes. For this requirement the longest possible wavelength, as in Figure 3.1, is equal to $2L$ where L is the distance between the anchor points. This mode is the fundamental vibration whose frequency v is defined by

$$v = \frac{v}{\lambda} = \frac{v}{2L} = \frac{1}{2L}\left(\frac{T}{m}\right)^{1/2} \tag{3.2}$$

Several of the higher modes are also shown in the figure, with wavelengths as indicated. The corresponding frequencies are harmonics of the fundamental frequency; thus the second harmonic occurs at $\lambda = L$, the third at $\lambda = 2L/3$ and so on. Only those wavelengths that are integer (commensurable) fractions of L can be accommodated.

The individual frequencies available to the system are, from Eq. (3.2) and the sense of Figure 3.1,

$$v_n = \frac{nv}{\lambda} = \frac{nv}{2L} = \frac{n}{2L}\left(\frac{T}{m}\right)^{1/2} \tag{3.3}$$

where n has the integer values 1, 2, 3, ... These resonant frequencies, each one corresponding to a particular value of n, are the characteristic values, or eigenvalues, of the function. The term *eigenvalue* thus denotes a simple idea and was used in classical physics prior to its introduction into quantum mechanics. Much of classical wave theory actually was developed by Rayleigh (1877) in his studies of sound. Looking again at Figure 3.1, we realize that if the following conditions were to exist:

Anchor points infinitely rigid
Vibrating system *in vacuo*
String perfectly elastic

the energy would remain confined to the vibrating cord (or to the space occupied by the vibrations) and the vibrations would continue indefinitely. In further generalization, these properties would hold for any oscillator, microscopic or macroscopic, if frictional damping were absent and radiation of energy did not occur.

Some of the properties of classical waves are analogous to those of quantum mechanics waves, and an understanding of the former will assist in a study of the latter. We should keep in mind the main points discussed or implied in the foregoing, which are:

1. Very often a vibrating mass, exemplified by the stretched cord in Figure 3.1, is the source of a wave. The energy of such a source is always the sum of the potential and kinetic forms and changes smoothly from one to the other during each cycle. It is entirely potential when the displacement is a maximum and the velocity is zero; it is entirely kinetic when the displacement is zero and the velocity is a maximum. A familiar example is seen in the swinging of a pendulum. In the background of any wave some kind of simple harmonic motion usually will be found.

2. In contrast to the traveling energy of a propagating wave, the energy of a standing wave remains localized in a volume of space determined by the system configuration and boundary conditions. If there is no damping, the wave persists indefinitely in this volume, and the resonant frequency is sharply defined. A useful measure of the frequency sharpness is the Q value of the oscillator, as pointed out in Section 1.2.3.

3. Even in a classical standing wave we see a frequency quantization in that only certain frequencies are permitted by the constraints. For a vibrating cord these are the length, tension, mass per unit length, and the boundary condition that the anchor points be vibration nodes.

4. A classical oscillating element, whether a short segment of a vibrating cord or the tip of a tuning fork, spends most of its time in the side regions where the displacement is greater and the velocity smaller. These are the regions near and bounded by the velocity reversal points. For example, the mass element of a simple harmonic oscillator spends two-thirds of the time at displacements greater than one-half of the maximum. Thus at any random instant the probability is greater that the element will be found near or at a reversal point than at the midpoint of the swing.

3.1.2. de Broglie Matter Waves

A balanced view of the nature of light suggests that the photons of quantum theory are guided in their paths by the electromagnetic waves of classical physics. This view expresses the principle of complementarity noted in Section 1.3.3. Following this trend of thought, de Broglie (1929) postulated in 1924 that a moving particle, an electron for example, has associated with it a companion wave called a *matter wave*, in addition to the familiar particle attributes of momentum and kinetic energy. He used this concept to explain the quantization of electron energy states in the hydrogen atom more satisfactorily than had been possible with the early Bohr model of the atom. Striking confirmation of matter waves came from the experiments by Davisson (1928) and Germer, who discovered that electrons scattered from a beam by a crystal create a diffraction pattern, indicating that some kind of waves do accompany electrons in their trajectories.

The de Broglie wave that accompanies the particle and travels at the same velocity is represented by its wave function Ψ. This function relates to the probability of finding the particle within a specified small volume and leads to the determining of its momentum and energy states. An early step in this direction was de Broglie's equation

$$\lambda = \frac{h}{mv} = \frac{h}{p} \tag{3.4}$$

where λ is the wavelength of the matter wave, m is the particle mass, and p is its momentum. The concept applies to large particles as well, but the resulting wavelengths are much too short to be in any observable category. This may be seen by considering two extreme cases. From Eq. (3.4) we find that an electron with mass $m = 9.1 \times 10^{-28}$ g and velocity $v = 100$ cm sec^{-1} is accompanied by a wave having $\lambda = 7.3 \times 10^{-2}$ cm. In contrast, the wavelength associated with a golf ball having mass 45 g and speed 2000 cm sec^{-1} has the value $\lambda = 7.2 \times 10^{-32}$ cm, which is less than that of the electron wave by a factor of 10^{-30}. This disparity is reminiscent of the great difference between quantized increments and total energy of a macroscopic oscillator, as discussed in Section 1.3.1.

The wave function for a particle moving in a $+x$ direction with constant momentum can be written as a sinusoidal function having amplitude A,

$$\Psi = A \sin(kx - \omega t) \tag{3.5}$$

which is an ordinary wave equation. However, without qualification this cannot represent a matter wave because it describes a wave of constant amplitude and unspecified duration. To appreciate this, keep in mind that the wave function Ψ is related to the probability of finding the particle at a particular place and time. We then expect that its wave representation should be a short group or packet of waves whose amplitude provides the clue to locating the particle. Such a wave group is illustrated by part (a) of Figure 3.2, where the waveform resembles that shown in Figure 1.6 for the packet accompanying a photon. If the de Broglie wave group is short, qualitatively we can take the center of the group as corresponding to the most probable location of the particle.

It becomes evident from Figure 3.2 that there is a basic difficulty in locating the particle accurately and simultaneously determining its momentum. This can be realized in a general way from the figure. In part (a) the maximum amplitude of the wave is well defined, hence the particle position can be pinpointed. However, the group is too short for establishing the wavelength, from which the momentum could be found according to Eq. (3.4). In part (b) the maximum is broader so that the particle position becomes uncertain, but the group is still too short for accurate determination of wavelength. In the extreme case of part (c) there is no maximum, as such, hence the particle position is diffuse, although the wavelength could now be determined to good accuracy. It can be shown (see

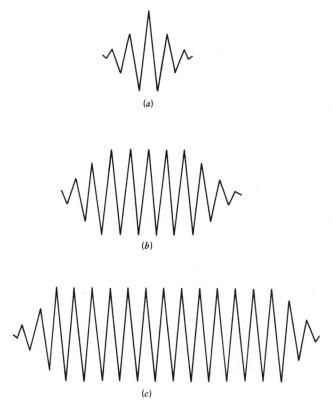

Figure 3.2. de Broglie wave groups accompanying a moving particle.

Beiser, 1967 or Eisberg and Resnick, 1974) that the position uncertainty Δx and the momentum uncertainty Δp are related by

$$\Delta x \, \Delta p \geq \frac{h}{2\pi}$$

which is the uncertainty principle given by Eq. (1.24).

Although the function Ψ has no easy physical interpretation, the quantity $|\Psi|^2$ when evaluated for a particular place and time is proportional to the probability of finding the particle there at that time. Actually, while Ψ is only *related* to the probability, $|\Psi|^2$ is a *probability density function* that, when normalized to unity, expresses a probability per unit volume. The product $|\Psi|^2 \Delta V$ gives the probability of finding the particle in a given differential volume ΔV, which may be interpreted as the volume element $dx \, dy \, dz$ centered on the point x, y, z. A central problem in quantum mechanics is to determine the function Ψ, hence the probability density $|\Psi|^2$, for specific cases where a particle is subject to external forces that limit its freedom of action. Such a particle may be one of the atomic nuclei making up a molecule or one of the electrons bound to an atom, vibrating with respect to, or revolving around, some reference point in the atom or molecule. We may regard the corresponding wave of the particle as a standing wave similar to those pictured in Figure 3.1.

3.1.3. Schroedinger Wave Equations

Quantum mechanics has developed around the Schroedinger wave equations in which the wave function Ψ is firmly embedded. The equation in *time-dependent* form and for one dimension is

$$\frac{h}{2\pi i} \frac{\partial \Psi}{\partial t} = \frac{h^2}{8\pi^2 m} \frac{\partial^2 \Psi}{\partial x^2} - \mathscr{E}_p \Psi \tag{3.6}$$

Here $\mathscr{E}_p$ is the potential energy of the particle, considered to be a point mass, as a function of position x and time t. In three-dimensional form the equation is

$$\frac{h}{2\pi i} \frac{\partial \Psi}{\partial t} = \frac{h^2}{8\pi^2 m} \left(\frac{\partial^2 \Psi}{\partial x^2} + \frac{\partial^2 \Psi}{\partial y^2} + \frac{\partial^2 \Psi}{\partial z^2} \right) - \mathscr{E}_p \Psi \tag{3.7}$$

In general, any restrictions on the particle's motions will affect its potential energy. Once $\mathscr{E}_p$ is known, solutions can be obtained from either equation for Ψ, from which the probability density can be determined for specified values of position coordinates and time.

In many cases relating to energy states, including those of interest here, the energy of the particle does not depend on time directly but on the position of the particle. When this condition exists, a simpler form of the Schroedinger equation can be reached by separating the variables x and t. To do this we can employ a standard technique of solving differential equations. This is to search for solutions that are the products of functions each of which contains only one of the independent variables involved in the equation. The technique reduces the partial differential equation to a set of ordinary differential equations, producing in this instance a simplification of the basic Schroedinger equation. Hence we write the wave function Ψ as the product of two variables,

$$\Psi(x, t) = \psi(x)\phi(t) \tag{3.8}$$

where ψ is a function of x alone and ϕ is a function of t alone. This lower case ψ must be carefully distinguished from the capital Ψ.

Dealing with the one-dimensional form (3.6), and substituting (3.8) therein, we get

$$\frac{h^2}{8\pi^2 m} \frac{\partial^2 [\psi(x)\phi(t)]}{\partial x^2} - \mathscr{E}_p \psi(x)\phi(t) = \frac{h}{2\pi i} \frac{\partial [\psi(x)\phi(t)]}{\partial t} \tag{3.9}$$

Here the potential energy $\mathscr{E}_p(x)$ is taken as a function of position only, in accordance with the previous statement. Performing the indicated differentiations leads to

$$\frac{h^2}{8\pi^2 m} \left(\frac{d^2\psi}{dx^2}\right) \phi(t) - \mathscr{E}_p(x)\psi(x)\phi(t) = \frac{h}{2\pi i} \frac{d\phi}{dt}\psi(x) \tag{3.10}$$

Dividing both sides by $\psi(x)\phi(t)$ gives

$$\frac{h^2}{8\pi^2 m} \frac{1}{\psi} \frac{d^2\psi}{dx^2} - \mathscr{E}_p = \frac{h}{2\pi i} \frac{1}{\phi} \frac{d\phi}{dt} \tag{3.11}$$

The left-hand side is a function of x only, and the right-hand side is a function of t only. Since the two sides are equal, it follows that each must be equal to the same constant. This constant has the dimensions of energy ML^2T^{-2} (which the reader may wish to verify) and indeed is identified with the total energy of the particle, as shown in more-detailed treatments.

Denoting this constant by $-\mathscr{E}$ we then get two equations from Eq. (3.11), namely,

$$\frac{h^2}{8\pi^2 m} \frac{d^2\psi}{dx^2} + (\mathscr{E} - \mathscr{E}_p)\psi = 0 \tag{3.12}$$

and

$$\frac{d\phi}{dt} = -\frac{2\pi i}{h}\mathscr{E}\phi \tag{3.13}$$

Equation (3.12) will be the principal concern, while Eq. (3.13) is used only in the following paragraph. In three-dimensional form, Eq. (3.12) is

$$\frac{\partial^2\psi}{\partial x^2} + \frac{\partial^2\psi}{\partial y^2} + \frac{\partial^2\psi}{\partial z^2} + \frac{8\pi^2 m}{h^2}(\mathscr{E} - \mathscr{E}_p)\psi = 0 \tag{3.14}$$

Equations (3.12) and (3.14) are the *time-independent* or steady-state forms of Schroedinger's equations. The requirements on each one are that the solution ψ and its derivatives everywhere be singlevalued, continuous, and finite. For atomic and molecular entities these expressions replace the equations of motion that underlie classical mechanics.

At this point it is convenient to develop the relationship between the functions $\Psi(x, t)$ and $\phi(t)$ beyond the symbolism of Eq. (3.8). To do this we need to express $\phi(t)$ in terms of its parameters. Multiplying both sides of Eq. (3.13) by dt/ϕ and integrating, we obtain

$$\ln\phi = -\frac{2\pi i}{h}\mathscr{E}t + \ln\phi_0 \tag{3.15}$$

The last term is a constant of integration which, for our purposes, can be ignored. Equation (3.15) then can be written as

$$\phi = \exp(-2\pi i\mathscr{E}t/h)$$

Substituting this into Eq. (3.8) gives

$$\Psi(x,t) = \psi(x)\exp(-2\pi i\mathscr{E}t/h) \tag{3.16}$$

The quantity $\psi^2(x)$ when normalized has an interpretation equivalent to the probability density $|\Psi|^2$ discussed in Section 3.1.2. That is, $\psi^2(x)$ is the probability density per unit length, and the product $\psi^2(x)dx$ is *equal* to the probability of finding the particle between x and $x + dx$.

The full meaning of the Schroedinger equations, however, goes beyond the probability aspects. For example, solutions of Eqs. (3.12) and (3.14) lead to certain *discrete* values of the total energy $\mathscr{E}$ for the time-independent or stationary states. As a rough analogy from classical mechanics, Eq. (3.3) of a vibrating cord is solvable only for certain frequencies as dictated by the integer n. There the frequency range is not continuous but is effectively quantized. Here the permissible energy values $\mathscr{E}_n$ are quantized by the number n; they are the *eigenvalues* of the wave equation. Complete solutions also provide the corresponding *eigenfunctions* ψ. These functions have certain integer numbers of nodes, where the integers are the *quantum numbers* for rotational, vibrational, and electronic energies. A preliminary example for an oscillator was cited without elaboration in Eq. (1.12).

3.1.4. Particle Trapped in a Well

In this section the time-independent Schroedinger equation for one dimension is considered, while in Section 3.2.1 the equation for three dimensions is applied to the enumeration of energy states. An example frequently used to illustrate the one-dimensional case is the particle in a "square well" of potential energy. This is often called *the particle in a box*. Referring to Figure 3.3 the particle is traveling back and forth along the x axis in a region bounded by walls at $x = 0$ and $x = L$. The potential energy between the walls is assumed to be zero, hence the total energy of the particle is kinetic. The particle and walls are assumed to be infinitely hard so that each collision with a wall is perfectly elastic and the particle does not lose energy by the encounter. The kinetic energy of the particle therefore remains constant. The potential energy becomes infinite at the walls and remains so outside the walls, hence the term *deep well*. As a result the particle is trapped. Here we have a crude, one-dimensional analogy of an electron bound to a small region of space within an atom where the relevant dimensions typically may be about 0.1 nm.

For the above assumptions, recalling that the potential energy is zero within the well and denoting the kinetic energy by $\mathscr{E}$ without subscript for simplicity, Eq. (3.12) becomes

$$\frac{d^2\psi}{dx^2} + \frac{8\pi^2 m}{h^2}\,\mathscr{E}\,\psi = 0 \tag{3.17}$$

The two solutions of this equation are

$$\psi = A \, \cos\left(\frac{8\pi^2 m \mathscr{E}}{h^2}\right)^{1/2} x$$

$$\tag{3.18}$$

$$\psi = B \, \sin\left(\frac{8\pi^2 m \mathscr{E}}{h^2}\right)^{1/2} x$$

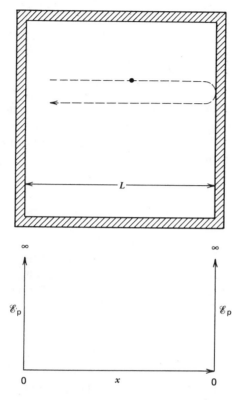

Figure 3.3. Particle trapped in a deep well of potential energy. From Beiser (1967). *Concepts of Modern Physics.* © 1963 by McGraw-Hill, Inc. Used with the permission of McGraw-Hill Book Co.

where A and B are constants. The solutions are subject to the boundary conditions $\psi = 0$ for $x = 0$ and $\psi = 0$ for $x = L$, because a node of the matter wave must occur at each wall, as with the vibrating cord in Figure 3.1. The first solution then must be rejected because ψ cannot be zero at $x = 0$ unless $A = 0$, which would render the proceedings nugatory. Thus any value for A is immaterial, but a good value for B in the second solution is readily found. The second solution allows B to be zero at $x = 0$, as required, but will be equal to zero at $x = L$ only when

$$\left(\frac{8\pi^2 m \mathscr{E}}{h^2}\right)^{1/2} L = n\pi, \qquad n = 1, 2, 3, \ldots \qquad (3.19)$$

where the values of n (always positive integers) are the *quantum numbers*.

The square of Eq. (3.19) gives the permissible values of energy $\mathscr{E}_n$, which are

$$\mathscr{E}_n = \frac{n^2 h^2}{8mL^2}, \qquad n = 1, 2, 3, \ldots \tag{3.20}$$

The values of $\mathscr{E}_n$, which are all discrete, are the energy states of the particle.

We next consider the wave function Eq. (3.18) from which Eq. (3.20) was derived to get the energy values. Substituting the value of $\mathscr{E}$ from the latter equation into the former, we have

$$\psi_n = B \, \sin\left(\frac{\pi n x}{L}\right) \tag{3.21}$$

The integral over all space of this expression squared is proportional to the probability of finding the particle at a certain position x. However, "all space" for the particle trapped in the well and moving along the x axis is merely the distance between $x = 0$ and $x = L$, as in Figure 3.3. Carrying out this integration for the square of Eq. (3.21) we find

$$\int_{-\infty}^{\infty} \psi n^2 \, dx = \int_0^L \psi_n \, dx = B^2 \frac{L}{2} \tag{3.22}$$

The quantity ψn^2 can be made *equal* to the probability $\mathscr{P}$ of finding the particle at position x by normalizing it so that

$$\int_0^L \psi n^2 \, dx = \int_0^L \mathscr{P} \, dx = 1 \tag{3.23}$$

which says, in effect, that the particle must be *somewhere* along the x axis between the limits 0 and L. Looking at Eq. (3.22) in the light of Eq. (3.23) it is clear that the former will be normalized to unity by using the value

$$B = \left(\frac{2}{L}\right)^{1/2} \tag{3.24}$$

for the constant in Eqs. (3.18) and (3.21).

Hence the normalized function is, from Eqs. (3.21) and (3.24),

$$\psi_n = \left(\frac{2}{L}\right)^{1/2} \sin\left(\frac{\pi n x}{L}\right) \tag{3.25}$$

This is plotted in Figure 3.4 for the quantum numbers 1, 2, and 3. It is seen that only the wave functions for which n is an integer will fit between the walls of infinite potential energy which lie at $x = 0$ and $x = L$. Here the analogy with the standing waves of a vibrating cord anchored at each end, as discussed in Section 3.1.1, is close. This is usually a helpful view of the Schroedinger equation, in whatever form it may be cast. In part (b) of the figure are plotted the corresponding probability density functions $\psi_n{}^2$, where the total area under each curve is unity according to Eq. (3.23). Interestingly, the probability of the particle being found at a particular value of x varies greatly with the quantum number. For example, the probability of being at the center of the well is a maximum for $n = 1$, while it is zero for $n = 2$. In contrast, classical physics would predict a finite probability for the moving particle to be found anywhere in the well. Any conceptual difficulty here can be relieved by recalling that the particle is being treated as a matter *wave*. In such a context nodes and antinodes are to be expected.

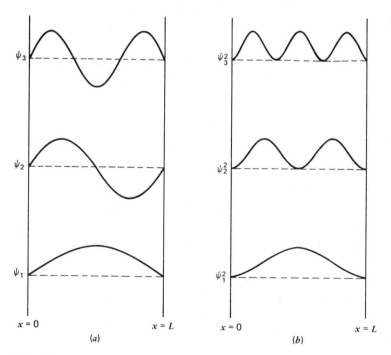

Figure 3.4. (a) Wave functions and (b) probability densities for a particle in a deep well of potential. From Beiser (1967). *Concepts of Modern Physics.* © 1963 by McGraw-Hill, Inc. Used with the permission of McGraw-Hill Book Co.

A more-realistic one-dimensional analogy to an electron bound within an atom is provided by the square well having low sides, shown in Figure 3.5. The particle is now in a shallow well of potential energy, and the resulting wave function has characteristics different from those just considered. Here the potential energy has finite values at the walls where $x = 0$ and $x = L$. The walls are now regarded as somewhat soft rather than infinitely rigid; the corresponding situation for a vibrating cord would involve slightly movable anchor points at the walls. Now the potential energy $\mathscr{E}_p$ has values relative to the kinetic energy $\mathscr{E}_k$ of the particle such that $\mathscr{E}_p > \mathscr{E}_k$ in regions I and III and $\mathscr{E}_p < \mathscr{E}_k$ in Region II.

The shallow well is more difficult to treat than the deep well (see Eisberg and Resnick, 1974 or Sproull, 1963) but some of the important results may be seen in Figure 3.6. In part (a) are plotted the wave functions ψ for $n = 1$, 2, and 3, while the corresponding probability densities ψ^2 are shown in part (b). Because ψ^2 does not quite equal zero at the walls, the wavelengths that

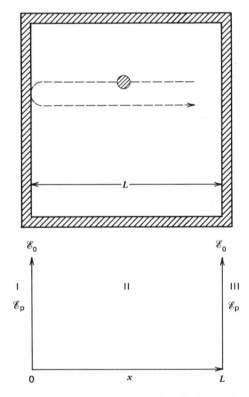

Figure 3.5. Particle trapped in a shallow well of potential energy.

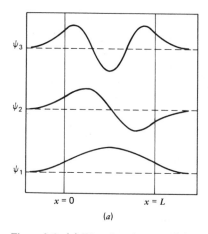
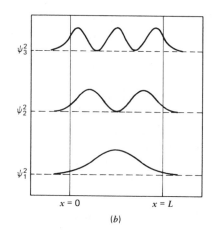

$x = 0$ $x = L$ $x = 0$ $x = L$

(a)   (b)

Figure 3.6. (*a*) Wave functions and (*b*) probability densities for a particle in a shallow well of potential. From Beiser (1967). *Concepts of Modern Physics.* © 1963 by McGraw-Hill Inc. Used with the permission of McGraw-Hill Book Co.

can be fitted into the well are somewhat longer than in the deep-well case. The longer waves mean lesser particle momenta and energies for the same value of n. Because ψ^2 is not zero at the walls, there is now a probability, albeit a small one, that the particle may be found outside the well. Since strictly $\mathcal{E}_k < \mathcal{E}_p$, such penetration of the wall is forbidden classically. Again, however, the particle is regarded here as a wave, and this permits an analogy from optics. A light wave undergoing reflection at a media interface actually penetrates the second medium a few wavelengths before reversing direction. Here the nonzero values of ψ and ψ^2 at the walls, as in Figure 3.6, portend the behavior of a harmonic oscillator confined to a small region of space.

3.2. QUANTUM STATES AND ENERGIES

The quantizations of atomic–molecular energies into discrete states, and the ways in which the particles can be distributed over these states, are basic in understanding absorption–emission. From the viewpoint of quantum mechanics there exists a very large number of unique energy states available to each particle. The energy values of the states are discrete, although the values may differ by only small increments and may be equal in certain groupings. A particle can change states rapidly as a result of collisions or by absorption–emission, but can be in only one state at any one time. In this section we examine the nature of energy states and their definition in terms of quantum numbers. This leads to the concept called *density of quantum states*, that is, the number of accessible states within a given volume. We then consider the grouping of states into energy levels and the resulting statistical weights of the levels. These ideas, as developed in later chapters, enable us to determine the average number of molecules occupying any particular level, for a given temperature. From these microscopic views of system energy we then define the macrostate of the system and its entropy. In conclusion we review the statistical interpretation of entropy and see that entropy is the link between classical thermodynamics and the microscopic view of matter.

3.2.1. Definition and Enumeration of Quantum States

Associated with each energy state of a particle is a small but determinable amount of space. This association is indicated by the spatial parameter L^2 in Eq. (3.20). Also, the wave function ψ is tied to L in Eq. (3.25). We therefore inquire into the number of energy states—considered here in the most general sense—that are accessible to a particle in a given volume of space. This is done by returning to the particle in a deep well or box and extending the freedom of particle motion to the y and z axes. The box is now three dimensional, and we assume that one of its corners is located at the origin of coordinate axes xyz with the edges extending in the positive directions. As in the simpler case, the potential energy is zero within the box and becomes infinite at each face. The energy state equation (3.20) is now extended to read

$$\mathscr{E}(n_x,\, n_y,\, n_z) = \frac{h^2}{8m} \left(\frac{n_x^2}{x^2} + \frac{n_y^2}{y^2} + \frac{n_z^2}{z^2} \right) \qquad (3.26)$$

where n_x, n_y, and n_z are the quantum numbers (positive integers) for the corresponding axes. The terms x, y, and z are the edge dimensions of the box.

Equation (3.26) is expressed more concisely by

$$\mathscr{E}_i = \frac{h^2}{8m} \frac{n_i^2}{L^2} \qquad (3.27)$$

where n_i collectively represents the three quantum numbers according to

$$n_i^2 = n_x^2 + n_y^2 + n_z^2 \qquad (3.28)$$

for any ith state. Also in this three-dimensional context, the length quantity is defined by

$$L^2 = x^2 + y^2 + z^2 \tag{3.29}$$

A change of unity in any of the three quantum numbers represented by n_i creates a different energy state. The values of $\mathcal{E}_i$ depend only on the values of n_i^2 and not on the individual n_x, n_y, n_z. From the classical viewpoint this can be interpreted as meaning that the energy state depends on the magnitude of the particle momentum but not on the direction.

For simplicity we assume that the box containing the particle is a cube of edge length L, so that its volume V equals L^3. Then L^2 equals $V^{2/3}$ and Eq. (3.27) can be written as

$$\mathcal{E}_i = \frac{h^2}{8m} \frac{n_i^2}{V^{2/3}} \tag{3.30}$$

for the energy of any ith state. Hence for a given energy value the representation n_i of the three quantum numbers must increase as the volume is made larger. Numerical examples of this relationship are cited in the following discussion. We next determine how many of these states have energy values equal to or less than some reference value to be selected later. That is, the energies of these states must conform to $\mathcal{E}_i \leq \mathcal{E}_0$. From Eq. (3.30) this requires the condition

$$n_i \leq 2 \frac{V^{1/3}}{h} (2m\mathcal{E}_0)^{1/2} \tag{3.31}$$

which reduces the problem to one of finding the numerical value of n_i for given values of $\mathcal{E}_0$ and V.

To do this we consider the octant of *quantum number space* shown in Figure 3.7. The restriction to an octant is due to the quantum numbers being positive integers. In principle this quantum number space is not unlike the momentum space employed in the discussion of classical entropy in Section 2.2.5. Here the complete set of all possible combinations of the quantum numbers according to Eq. (3.28) forms a cubical lattice, which consists of points whose coordinates are integer values of these numbers. Within the interior of the lattice there is one cube of unit quantum number volume for each lattice point, as suggested by the figure. Each of these little cubes (cells) represents just one discrete energy state. Equation (3.31) shows that all of these states must lie within the octant whose radius, in units of quantum number n, is defined by that expression. If the macroscopic physical volume is held constant, the number of energy states depends only on the value of n_i.

From the geometry of Figure 3.7 it is seen that all states of equal energy must lie on the spherical surface whose radius is n_i. This surface will theoretically cut through some of the cells, so it is uncertain whether a point representing a given energy state is just inside, or just outside, the surface. However, when the quantum numbers are large, which is true for atmospheric molecules, the uncertainty becomes very small relative to the total number of states. This total number then is about equal to the volume of the octant, expressed in units of n. Hence the total number Γ of states with energy less than $\mathscr{E}_0$ is

$$\Gamma = \frac{1}{8} \frac{4\pi}{3} n_i^3 = \frac{\pi}{6} n_i^3$$

Substituting for n_i its value from the requirement Eq. (3.31), we get

$$\Gamma = \frac{\pi}{6} \left[2 \frac{V^{1/3}}{h} (2m\mathscr{E}_0)^{1/2} \right]^3$$

$$= \frac{4\pi}{3} \frac{V}{h^3} (2m\mathscr{E}_0)^{3/2} \tag{3.32}$$

Let us next assume that the reference value of $\mathscr{E}_0$ is the average kinetic energy of a molecule given by Eq. (2.45) as $3\kappa T/2$. At 0°C this has the value 5.658×10^{-21} J. Taking for an example a volume $V = 1$ m^3 and an O_2 molecule having $m = 5.31 \times 10^{-26}$ kg, we find from Eq. (3.32),

$$\Gamma = 2.1 \times 10^{32} \text{ accessible states} \tag{3.32a}$$

which have energy values equal to or less than $3\kappa T/2$ at 0°C. Such a result is independent of the shape of the volume provided that its dimensions are large compared to the wavelength of the matter wave defined by Eq. (3.4). Since the value of $3\kappa T/2$ is not great, the large number of accessible states means that they must be closely spaced relative to the absolute values of their energies. Stated differently, the density of states in quantum number space is quite high.

For comparison with this number of quantum states, we derive from Eq. (2.7) that the number of molecules in 1 m^3 of gas at STP conditions is about 2.7×10^{25}. Even if all these molecules were evenly distributed over the 2.1×10^{32} states as found above (which would be highly improbable), many

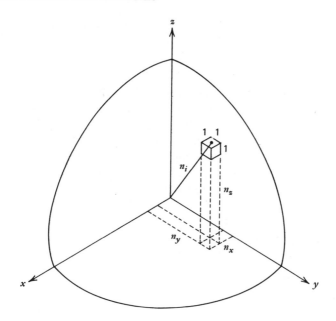

Figure 3.7. An octant of quantum number space. Each dot represents a particular combination of integer values of n_x, n_y, n_z (i. e., a particular quantum state) and thus occupies the center of a cube of unit volume. From Vincenti and Kruger (1965).

of the states would be empty. Only about 1 in every 10^7 states would be occupied. The quantum numbers for most of the accessible states, however, are quite high. This can be appreciated by solving Eq. (3.30) for the conditions used to get a numerical value from Eq. (3.32). The result is

$$n_i^2 \approx 5.5 \times 10^{21}$$

for an approximate upper limit. As a reminder, n_i represents the individual quantum numbers according to Eq. (3.28). From this relationship it is evident that many combinations of the numbers can be formed to produce a given large value of n_i^2.

Actually the *product sets* of quantum numbers, rather than the combinations, are significant because the order in which the numbers are assigned to the axes is important. Each product set corresponds to a particular quantum state. Such product sets are similar to permutations of the numbers but the factorial element does not apply, as is shown in the following. Assume that the same maximum quantum number n_i is valid for each axis.

Then for each value of n_x up to n_i there are n_i values of n_y, and for each of these there are n_i values of n_z. That is,

$$\text{Total number of sets} = n_{x,i} \times n_{y,i} \times n_{z,i} = n_i{}^3$$

Such sets of the digits 1, 2, 3 are listed in Table 3.1, along with the resulting values of $n_i{}^2$, which would lead to $\mathscr{E}_i$ by way of Eq. (3.30). A total of 27 distinct products exists even for the modest value $n_i = 3$, and each product represents a separate state. For usual situations where n_i is large the possibilities are nearly endless. It is seen in the table that a given $n_i{}^2$, hence a particular state $\mathscr{E}_i$, is produced by several sets of products of the individual quantum numbers. These exemplify the multiplicity of states, to which we now turn.

TABLE 3.1 PRODUCT SETS OF THE FIRST THREE QUANTUM NUMBERS

n_x	n_y	n_z	$n_i{}^2$
1	1	1	3
2	1	1	6
3	1	1	11
1	2	1	6
2	2	1	9
3	2	1	14
1	3	1	11
2	3	1	14
3	3	1	19
1	1	2	6
2	1	2	9
3	1	2	14
1	2	2	9
2	2	2	12
3	2	2	17
1	3	2	14
2	3	2	17
3	3	2	22
1	1	3	11
2	1	3	14
3	1	3	19
1	2	3	14
2	2	3	17
3	2	3	22
1	3	3	19
2	3	3	22
3	3	3	27

3.2.2. Energy States, Levels, and Degeneracies

We introduce here the ideas of quantum or energy states grouped into energy levels, and also the degeneracies and statistical weights of the levels. An energy level is defined as a group of states having identical values. Sometimes it is helpful to regard the level as consisting of states that have *very nearly* the same values, where the differences are extremely small compared to the absolute values. This point of view is employed in deriving expressions for the densities of states. When a level contains more than one state, it is called *degenerate* (not a moral judgment). The number of states in a level i is the degeneracy g_i of that level.

The *statistical weight,* also denoted by g_i, is numerically equal to the degeneracy and is a factor that must be applied in determining a population distribution. For example, if there are N_1 states in level g_1 and N_2 states in level g_2, the *a priori* probability of the population ratio of the two levels is

$$\frac{N_1}{N_2} \propto \frac{g_1}{g_2} \tag{3.33}$$

if no other factors were operative. This can be seen by considering Figure 3.8, which shows six horizontal layers of openings, or bins, in an otherwise impenetrable wall. The bins correspond to energy levels. Each bin is variously divided by vertical partitions into cells which represent states. The number of cells at any level is equal to the degeneracy of that level. Hence the area of a bin is proportional to the degeneracy of the level. Let us suppose that ordinary baseballs are pitched randomly at the total open-

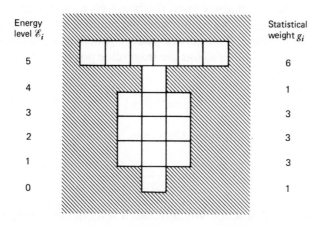

Figure 3.8. A physical analog of statistical weighting factor.

ing, and that only the total number N of balls that enter are counted. Thus far the setup is reminiscent of an amusement concession at a county fairground.

If the balls are indeed pitched randomly so that no part of the total opening is favored, after a large number of pitches the "density" of ball arrivals is uniform over the opening. Now let A be the area of the total opening; let g_0, g_1, $\ldots$, g_i be the areas of the bin openings at the various levels; and let a_0, a_1, $\ldots$, a_i be the areas of the cell openings. The probability that a ball will enter bin g_i is just equal to the ratio g_i/A, and the probability that it will enter cell a_i is equal to a_i/A. The catching ability of any opening is proportional to its area.

When N is large, the number of balls landing in bin g_i is, to a good approximation,

$$N_i = \frac{N}{A}\, g_i \tag{3.34}$$

Here we may regard the ratio N/A as the density of arrivals. Thus the area of the bin opening, which is equal to the sum of the areas of the component cell openings and hence is proportional to the degeneracy, is the statistical weighting factor for that level.

The ideas of energy levels, degeneracies, and populations are illustrated in Figure 3.9 by the shelves, boxes, and marbles, after the manner of Sears and Salinger (1975). Each shelf represents an energy level, while the number of boxes corresponds to the number of states at that level. Thus the degeneracy g_i of a level i is equal to the number of boxes on that shelf. The number of marbles in any one box is the number of particles in that state. The total number of marbles in all the boxes on a shelf is the *occupancy number* N_i of that level, no matter how the marbles are distributed.

Looking at Figure 3.9, the lowest-energy condition $\mathscr{E}_0$ exists when the resultant quantum number n_0 has the least value possible. From Eq. (3.28) this requires that

$$n_x = n_y = n_z = 1$$

which yields $n_0^2 = 3$. Equation (3.30) then becomes

$$\mathscr{E}_0 = \frac{3h^2}{8mV^{2/3}}$$

for the lowest or first level. We note that a level of zero energy does not exist. Because of the condition (3.28) with each quantum number equal to

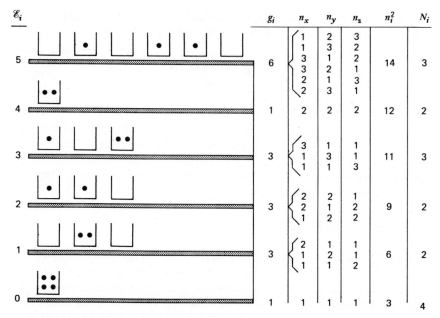

Figure 3.9. A physical model of energy levels, degeneracies, and populations. After Sears and Salinger (1975). *Introduction to Thermodynamics, Kinetic Theory, and Statistical Thermodynamics*, 3rd ed. © 1975 by Addison-Wesley, Reading, Mass. Used with permission.

unity, there is only one state at this level, as shown in the figure. Hence this level is not degenerate and the factor g_0 is unity. The four marbles in the single box on this shelf signify that the occupancy number $N_0 = 4$.

In the next level $\mathscr{E}_1$ upward, any one of the three quantum numbers can have the value 2 while the other two numbers are each unity. There are three possible groupings of the numbers: 211, 121, and 112. Each grouping, that is, each product set, corresponds to a different state. From Eq. (3.28) we see that $n_1^2 = 6$ for each set, so that Eq. (3.30) then becomes

$$\mathscr{E}_1 = \frac{3h^2}{4mV^{2/3}}$$

for this second level. Since it comprises three states having equal energies, this level has a degeneracy $g_1 = 3$, indicated by the three boxes. The two marbles can have various distributions among the boxes. Although only one distribution is shown, actually there are six identifiable arrangements that can be made.

In the third level $\mathscr{E}_2$, any two of the numbers can have the value 2 while the other number is unity. Now $n_2^2 = 9$, and Eq. (3.30) is

$$\mathscr{E}_2 = \frac{9h^2}{8mV^{2/3}}$$

The degeneracy evidently is $g_2 = 3$, identical to that of the second level. The two marbles, indicating that $N_2 = 2$, have six identifiable distributions among the three boxes. Also shown are the next three higher levels where $n_3^2 = 11$, $n_4^2 = 12$, and $n_5^2 = 14$, respectively. Similar calculations can be made for still higher levels, but the method is tedious and becomes impractical at large quantum numbers.

3.2.3. Density of Energy States

It is instructive to calculate the density of energy states and, in the process, gain a broad view of degeneracy. We define *state density* $G(n_i)$ as the number of quantum states per unit energy interval, per unit volume. We reach an expression for $G(n_i)$ by determining the number of states that have values between n_i and $n_i + \Delta n_i$. This implies a continuous function of n that would be inconsistent, strictly, with the integer values of n_x, n_y, n_z that make up n_i, as defined by Eq. (3.28). However, when these numbers are very large the unit-step increases of the numbers are relatively small, and the relationship can be approximated by a continuous function. Use of the differential Δn_i instead of dn_i means that the increases in question, although small, are finite.

The concept of state density can be visualized from the octant of quantum number space illustrated in Figure 3.7. In that connection the total number of states having energies equal to and less than a reference level $\mathscr{E}_0$ was counted and designated Γ. Now let that octant be enclosed, not by a spherical surface as in Figure 3.7, but by a spherical shell as shown in Figure 3.10. This shell, having radius n_i and thickness Δn_i, contains all the states whose energies correspond to n_i, or rather n_i^2 according to Eq. (3.30). The number of these states is equal to the volume of the shell expressed in units of quantum number. All of these states are in the same energy level because they all lie at the same radius from the center of the octant. Therefore, the degeneracy, which is equal to the number of states in the level, also is equal to the volume of the shell.

In terms of quantum number the volume of the spherical shell having radius n_i and thickness Δn_i, as in Figure 3.10, is given by

$$G_i = \frac{4\pi}{8} n_i^2 \Delta n_i = \frac{\pi}{2} n_i^2 \Delta n_i = g_i \qquad (3.35)$$

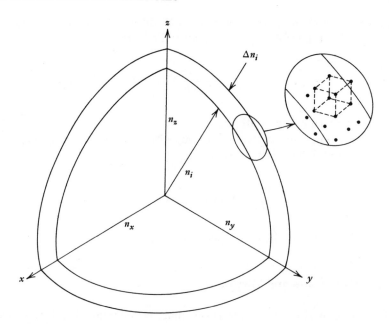

Figure 3.10. Density of energy states. Each dot in the octant represents a particular combination of integer values of n_x, n_y, and n_z, hence each dot represents a quantum state occupying a cube of unit volume. All states within the shell have approximately the same energy. From Sproull (1963).

which corresponds to the number of lattice points within the shell. It is seen that the degeneracy g_i of the level is defined by n_i when the number is very large, as in the present assumption. Because Δn_i has a finite rather than an infinitesimal value, we must regard the level, that is, the shell volume, as consisting of all states that have very nearly the same values of energy. This viewpoint was mentioned at the beginning of Section 3.2.2. The value of g_i, although overwhelmingly dependent on n_i, is also affected by the size of Δn_i, which thus acts like a grouping factor.

In terms of energy—a stronger interest here than quantum number itself —the degeneracy is found by changing the variable in Eq. (3.35). Substituting therein the value of $\mathscr{E}_i$ and $\Delta\mathscr{E}_i$ for n_i and Δn_i, as determined from Eq. (3.30), we get

$$G_i = \frac{4\pi}{h^3} Vm(2m\mathscr{E}_i)^{1/2}\Delta\mathscr{E}_i = g_i \qquad (3.36)$$

for the number of states having energies between $\mathscr{E}_i$ and $\mathscr{E}_i + \Delta\mathscr{E}_i$. An equivalent result can be obtained directly by differentiating Eq. (3.32), the expression for the total number of states. This yields

$$\frac{d\Gamma}{d\mathscr{E}} = G(\mathscr{E}, V) = \frac{4\pi}{h^3} V m (2m\mathscr{E})^{1/2}$$

for the density of energy states per unit energy interval, in a given volume. This agrees with the expression (3.36) obtained from the geometrical method. Dividing by the volume V we get

$$\frac{d\Gamma}{d\mathscr{E}} = G(\mathscr{E}) = \frac{4\pi}{h^3} (2m\mathscr{E})^{1/2} \tag{3.37}$$

for the density of energy states per unit volume.

3.2.4. Macrostates and Microstates

We distinguish between *macrostates* (macroscopic states) and *microstates* (microscopic states). From the viewpoint of the gas laws and thermodynamics the macrostate of a system is defined by its gross properties such as temperature, specific volume, and pressure. Statistical mechanics provides a microscopic interpretation by defining the macrostate in terms of particles and their energy states and levels, for example, by specifying the occupancy number N_i of each energy level. Thus in Figure 3.9 a particular macrostate is defined by the set of N_i's listed there. A different set of numbers would mean a different macrostate. Two important restrictions on the macrostates of a closed system must be kept in mind. First, the total number of particles in all levels remains constant and is given by the summation

$$\sum_i N_i = N \tag{3.38}$$

over all levels. Secondly, the total system energy must remain constant. Because all particles in any level i must have the same energy $\mathscr{E}_i$, the total energy $\mathscr{E}$ of the particles in the level is $N_i \mathscr{E}_i$, and the total system energy is

$$\sum_i N_i \mathscr{E}_i = \mathscr{E} \tag{3.39}$$

over all levels.

The meaning of the restrictions (3.38) and (3.39) is illustrated in a simple way by Figure 3.11. Here five indistinguishable particles are to be distributed over six energy levels for a total energy $\mathscr{E} = 8$ units. That is,

$$\sum_i N_i = 5 \quad \text{and} \quad \sum_i N_i \mathscr{E}_i = 8$$

Six candidate distributions are shown, and, since the occupancy numbers of the levels are specified, each of these distributions constitutes a macrostate. It is seen that the first three distributions meet the requirement $\mathscr{E} = 8$ units but the remaining two do not. Actually there are 14 valid macrostates for this system, a number that the reader may wish to verify.

A microstate has many different microstates, depending on the degeneracies of the levels, as discussed in the two preceding sections. The only accessible, or possible, microstates, however, are those that can constitute a macrostate in accordance with Eqs. (3.38) and (3.39). All such accessible microstates have equal probabilities of occurrence, and the number W_i of these microstates is called the *thermodynamic probability* of the macrostate. We note that this is different from the usual meaning of probability. The total number Ω of accessible microstates of a *system* is given by a

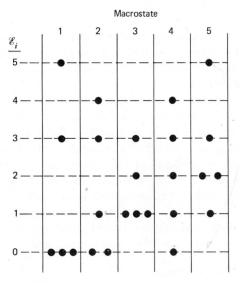

Figure 3.11. Some possible macrostates of a system having constant total population $N = 5$ particles and constant total energy $\mathscr{E} = 8$ units.

summation of the thermodynamic probabilities over all macrostates of the system,

$$\Omega = \sum_i \mathcal{W}_i \qquad (3.40)$$

where the numbers involved are very large.

A microstate is defined in either of two ways, depending on how the particles are regarded. If they are indistinguishable, a microstate is defined by stating the number of particles in each energy state of each level. Thus Figure 3.9, which corresponds to a macrostate, also depicts a microstate by showing the number of marbles in each of the boxes. However, a different microstate would be created if the three marbles in level $\mathcal{E}_3$ were reassigned among the three boxes, and this would be true for any other level. The macrostate would not be changed as long as the occupancy numbers did not change.

If the particles are distinguishable, the energy state of each particle must be specified to define the microstate. We must state which particle is in which state of each level. However, the order or sequence in which the distinguished particles are listed or arrayed in a given state is immaterial. An equal exchange of particles between states, in the same level or in different levels, creates new microstates. Such exchanges do not create a new macrostate if the occupancy numbers of the levels do not change. We may expect that the number of microstates is much greater for distinguishable than for indistinguishable particles.

In a gas system the molecular collisions and radiative transitions produce exchanges of energy that continually change the arrangements of particles with respect to energy states. The system thus passes through many microstates in short order, but, if radiative equilibrium exists, each macrostate is subject to Eqs. (3.38) and (3.39). Because all accessible microstates have equal probabilities, the system will rapidly explore, without guidance hence without favoritism, all of these microstates. The length of time that the system spends in the corresponding macrostate evidently is proportional to the number $\mathcal{W}_i$ of microstates in that macrostate. Therefore, the most probable macrostate is the one where $\mathcal{W}_i$ is a maximum, and this is where the system will be found. The maximum is so sharp that even small departures from this macrostate are so rare as to be practically nonexistent. This is the equilibrium condition toward which an isolated system irreversibly moves, regardless of initial conditions, and in which it remains.

3.2.5. Entropy and Information

In thermodynamics the concept of entropy is closely associated with the second law and establishes the direction in which heat will flow from one system to another. The heat always flows in that direction along which the entropy increases, and, if the process is irreversible, entropy establishes the direction in which time's arrow points. Statistical mechanics provides insight into these matters by considering entropy to be a quantitative measure of the *disorder* of a system. Boltzmann showed that the entropy S of an isolated system is proportional to $\ln \Omega$ and found that

$$S = \kappa \ln \Omega = \kappa \ln \sum_i \mathcal{W}_i \qquad (3.41)$$

where κ is the constant defined by Eqs. (2.45) and (2.47). Equation (3.41) links statistical mechanics to traditional thermodynamics by providing a definition of entropy which is additional to the one given by Eq. (2.26).

The idea of disorder has a subjective connotation which is difficult to eliminate entirely. Clearly our ideas of order in a system relate to how well its pieces fit certain patterns of position or behavior which are fixed as criteria. If we know that the fits are good, we say that the system is in order. For example, a data filing system is in order if each index card is in its proper alphabetical place. Suppose now that the cards are removed from the trays, thoroughly shuffled, and replaced. The filing system would now be considered as in fairly complete disorder, but—and this is the usually ignored fact—the new sequence of the cards would still constitute a pattern. We assert *disorder* but this is mostly because we do not know the location of any particular card. Thus in a strong sense the disorder merely reflects our ignorance.

Such disorder, however, is only with respect to the ordained alphabetical sequence which we knew in advance. The actual sequence in which the shuffled cards lie is just as much a pattern as is the alphabetical sequence. Did we but know the shuffled sequence as well as we know the alphabet, we could predict the location of a particular card to the same degree of probability as we now enjoy with a well-managed filing system. The probability that we assign to an event is related to the knowledge that we have, as in the quotation from Eddington at the beginning of this chapter. Shuffling the index cards destroyed the old order and created a new one, but the new order is not different in kind from the old one, although it may be less useful. From the viewpoint of our knowledge, mixing up the cards created disorder that further shuffling will not reduce. But, as suggested above, the disorder lies in the fact that we do not know the new arrangement. That is, the disorder we perceive is a confession of ignorance, and this is the connection between entropy and information theory. Readers wishing to pursue this further will find Pierce (1961) quite stimulating.

A particular example must suffice here. If we arrange a system so that only one microstate is available to it, then from Eq. (3.40) we have $\Omega = 1$

and from Eq. (3.41) the entropy $S = 0$. The system is in perfect order because the state of each particle can be specified exactly. This corresponds to the example of the ordered crystal at the absolute zero of temperature, as noted under the third law in Section 2.2.3. As we make more microstates available to our system, say by the addition of heat, the number Ω becomes larger. Now the energy state of each particle cannot be specified exactly, but only to some probability, because all of the microstates have equal probabilities. In this sense the disorder and the entropy have increased. In defining the system energy, the reliance must now be on statistical methods, as treated in the remainder of this chapter.

3.3. POPULATION STATISTICS AND DISTRIBUTIONS

With the ideas of quantum states and their energies in mind, we now examine the ways in which a molecular population can be distributed over large numbers of states and levels. This requires the methods of statistical mechanics, which pays no attention to an individual molecule but concentrates on the group behavior of large populations. Such behavior is manifested in three major distributions: Bose–Einstein, Fermi–Dirac, and Boltzmann. To each of these there corresponds a set of statistics that differ from each other only in initial assumptions regarding the particles. The differences seem minor at a cursory glance, but they are responsible for widely divergent distributions. For these reasons we first look at the statistics.

In dealing with the statistics of particle distributions and seeking molecular explanations of gas entropy, classical physics regarded the particles as *distinguishable*. The Boltzmann statistics rests upon this postulate. As quantum theory advanced, so did the necessity for regarding the particles as *indistinguishable*. Particles of a given kind bear no labels—have no characteristics—that enable them to be separately identified. The Bose–Einstein and Fermi–Dirac sets of statistics rest upon this postulate. The first set was developed during studies of photons in a blackbody cavity, but is also applicable to the molecules of a gas. The second set of statistics came from studies of the electronic structure of atoms and embodies the Pauli exclusion principle. In dealing with the Boltzmann, Bose–Einstein, and Fermi–Dirac statistics, the first problem is to account for the possible arrangements of particles with respect to states and levels. The second problem is to find the most probable arrangements or distributions. We now assume an isolated system in which the total number of particles and the total energy remain constant, according to Eqs. (3.38) and (3.39), and examine the three sets of statistics.

3.3.1. Bose–Einstein and Fermi–Dirac Statistics

Bose–Einstein statistics are considered first. The particles are indistinguishable, and there are no restrictions on the number of particles that can be in any one state. Referring to the example in Figure 3.12, the particles 1, 2, 3, ..., N_i are lined up on an energy level, but are variously separated into g_i states (cells) by $g_i - 1$ partitions. The term g_i indicates the degeneracy or statistical weight of that level. Assume for a moment that the N_i particles and $g_i - 1$ partitions are distinguishable. Then the number of distinct ways in which the particles and partitions can be arranged, that is, the number of *permutations*, is equal to $(N_i + g_i - 1)!$. For the nine particles and four states (three partitions) in the figure we have

$$(9 + 4 - 1)! = 12! \approx 4.79 \times 10^8$$

permutations. This is a large number for so few participants.

However, the assumption that the $g_i - 1$ partitions are distinguishable is incorrect and has led to $(g_i - 1)!$ too many arrangements. Therefore the first result must be divided by $(g_i - 1)!$. In a similar vein the assumption that the $N!$ particles are distinguishable is incorrect and has led to $N!$ too many arrangements. A correction factor (division) of this amount is required. Hence we have

$$w_i = \frac{(N_i + g_i - 1)!}{(g_i - 1)! \, N_i!} \tag{3.42}$$

for the number of ways in which N_i particles can be distributed among g_i states in any ith level. Returning to the nine particles and four states of Figure 3.12 for an example, substitutions in Eq. (3.42) give

$$\frac{(9 + 4 - 1)!}{(4 - 1)! \, 9!} = 220 \text{ permutations}$$

which constrasts strongly with the first result.

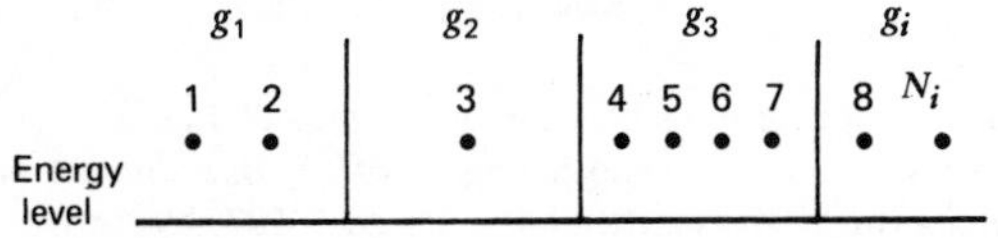

Figure 3.12. Particles and states for Bose–Einstein statistics. A series of N_i indistinguishable particles are separated into g_i cells (states) by $g_i - 1$ partitions.

The number of microstates given by Eq. (3.42) for any one level is independent of possible arrangements in other levels. The total number of microstates embracing all the occupied levels, or the thermodynamic probability of each macrostate, is the product over all such levels of the values given by Eq. (3.42) for each level. This is represented by

$$\mathcal{W}_j = \prod_i w_i = \prod_i \frac{(N_i + g_i - 1)!}{(g_i - 1)! \, N_i!} \tag{3.43}$$

where $\prod_i$ means that we should form the product of all terms following it, for each value of i. An example of such microstates and the resulting products is shown in Figure 3.13. The system assumed is the same as that

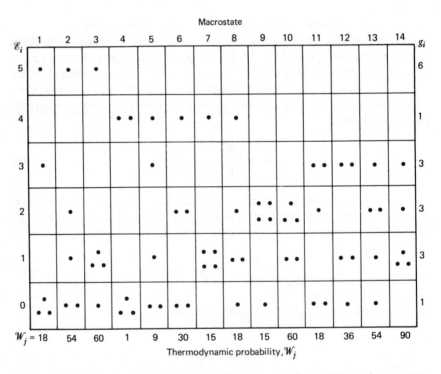

Figure 3.13. The 14 possible macrostates for the system of Figure 3.11 when governed by Bose–Einstein statistics. After Sears and Salinger (1975). *Introduction to Thermodynamics, Kinetic Theory, and Statistical Thermodynamics*, 3rd ed. © 1975 by Addison-Wesley, Reading, Mass. Used with permission.

used for Figure 3.11, namely, six equally spaced energy levels $\mathscr{E}_0$ through $\mathscr{E}_5$, total particle number $N = 5$, total energy $\mathscr{E} = 8$ units, and values of g_i as indicated. In Figure 3.13, however, all 14 possible macrostates are depicted, and the thermodynamic probability of each is listed beneath it. Each horizontal row represents an energy level, with its degeneracy indicated at the right. The marbles represent the number of particles in each level, as before, but no attempt has been made to show the states.

The values of $\mathscr{W}_j$ listed in Figure 3.13 are spread from unity for $j = 4$, to 90 for $j = 14$. All of the values were calculated from Eq. (3.43), and it is instructive to show several of these calculations. Taking macrostate 4, we have two particles in level $\mathscr{E}_4$, so that $N_4 = 2$, and $g_4 = 1$ for this level. The remaining three particles are in level $\mathscr{E}_0$, so that $N_0 = 3$, and $g_0 = 1$. Substituting in Eq. (3.43) for these two levels, which are the only occupied ones, we have

$$\mathscr{W}_4 = \frac{(2 + 1 - 1)!}{(1 - 1)!\,2!} \times \frac{(3 + 1 - 1)!}{(1 - 1)!\,3!} = \frac{2!\,3!}{2!\,3!} = 1$$

where $0! = 1$, by convention. Taking macrostate 10, where $N_2 = 3$, $g_2 = 3$, and $N_1 = 2$, $g_1 = 3$, and the other levels are inactive, substitutions give

$$\mathscr{W}_{10} = \frac{(3 + 3 - 1)!}{(3 - 1)!\,3!} \times \frac{(2 + 3 + 1)!}{(3 - 1)!\,2!} = \frac{5!\,4!}{2!\,3!\,2!\,2!} = 60$$

For the final example, we take macrostate 14, for which $N_3 = 1$ and $g_3 = 3$; $N_2 = 1$, $g_2 = 3$; and $N_1 = 3$, $g_1 = 3$. We get

$$\mathscr{W}_{14} = \frac{(1 + 3 - 1)!}{(3 - 1)!\,1!} \times \frac{(1 + 3 - 1)!}{(3 - 1)!\,1!} \times \frac{(3 + 3 - 1)!}{(3 - 1)!\,3!} = \frac{3!\,3!\,5!}{2!\,2!\,2!\,3!} = 90$$

Looking at the figure, all of these macrostates—and only these—are possible. As the system rapidly explores all of them, without favoritism, the amount of time that it spends in any one is proportional to the number $\mathscr{W}_j$ of microstates therein. Hence macrostate 4 is the least probable and macrostate 14 is the most probable.

In *Fermi–Dirac* statistics the particles are indistinguishable, but there can be no more than one particle in one state. This restriction is due to the Pauli exclusion principle, which asserts that no two electrons can be in any one quantum state. In effect, we have $N_i \leq g_i$. Thus if there are g_i states in an energy level and N_i particles in that level, then only N_i states can be occupied and $g_i - N_i$ are vacant. The g_i states themselves can be arranged (permuted) in $g_i!$ different ways. However, the $N_i!$ permutations of the N_i occupied states do not represent different arrangements because the particles are not distinguishable. Likewise, the $(g_i - N_i)!$ permutations of the vacant states are irrelevant because these states do not contain particles. Hence the number of distinct arrangements of particles and states is

$$w_i = \frac{g_i!}{N_i! \, (g_i - N_i)!} \qquad (3.44)$$

which represents the number of microstates in any ith level.

The number of microstates given by Eq. (3.44) for any one level is independent of possible arrangements in other levels. As in the Bose–Einstein statistics above, to find the number of microstates in any macrostate j we form the product $\mathcal{W}_j$, each of whose terms is Eq. (3.44) for each occupied level. This product is

$$\mathcal{W}_j = \prod_i w_i = \prod_i \frac{g_i!}{N_i! \, (g_i - N_i)!} \qquad (3.45)$$

where $\prod_i$ indicates that we are to form the products as we did previously. Figure 3.14 shows the possible macrostates permitted by the Fermi–Dirac statistics, employing the same system as used for the Bose–Einstein statistics in Figure 3.13. Because of the restriction that not more than one particle can occupy one state, however, the number of particles in any level cannot exceed the degeneracy of that level. Hence the Bose–Einstein macrostates 1, 2, 4, 5, 6, 7, 9, and 11 must now be disallowed, leaving only macrostates 3, 8, 10, 12, 13, and 14. These are shown in that order in Figure 3.14 but numbered anew from unity. The thermodynamic probability of each has been calculated from Eq. (3.45) and listed beneath the column.

A sample calculation from Fermi–Dirac statistics will illustrate the method. Taking macrostate 6 of Figure 3.14 for the three occupied states we have $N_3 = 1$, $g_3 = 3$; $N_2 = 1$, $g_2 = 3$; and $N_1 = 3$, $g_1 = 3$. Substituting into Eq. (3.45) we have

$$\mathcal{W}_j = \frac{3!}{1! \, (3 - 1)!} \times \frac{3!}{1!(3 - 1)!} \times \frac{3!}{3! \, (3 - 3)!} = 9$$

which contrasts with the value 90 for the corresponding macrostate 14 of Figure 3.13. Comparing the two figures we see, even from our miniature system, that the restriction of one particle per state severely limits the number of arrangements.

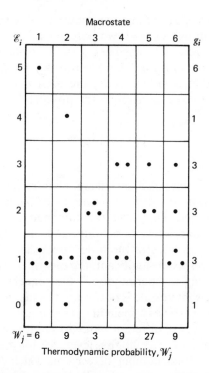

Figure 3.14. The six possible macrostates for the system of Figure 3.11 when governed by Fermi–Dirac statistics. After Sears and Salinger (1975). *Introduction to Thermodynamics, Kinetic Theory, and Statistical Thermodynamics,* 3rd ed. © 1975 by Addison-Wesley, Reading, Mass. Used with permission.

3.3.2. Boltzmann Statistics

In Boltzmann statistics the particles are regarded as distinguishable, and there are no restrictions on the number of particles in any one state. Consider now a system of N particles, and a macrostate whose energy levels have occupancy numbers N_1, N_2, N_3, ..., N_i, and degeneracies g_1, g_2, g_3 ..., g_i. Taking any level i, the first particle from the group N_i can be put into any one of the g_i states of this ith level. The second particle also can be put into any one of these states, and so on to the last particle from this group. Therefore, the number of possible distributions of the N_i particles among the g_i states of this level is

$$\mathcal{W} = g_i{}^{N_i} \tag{3.46}$$

Table 3.2 depicts these possibilities on a small scale for one level only. Here are four states g_1 through g_4, and two particles a and b to be distributed among the states. Sixteen arrangements or microstates are seen to be possible, in agreement with Eq. (3.46). Since the distribution in this level is independent of that in other levels, the total number $\mathcal{W}_i$ of possible distributions over all levels is given by the product

$$\prod_i w_i = \prod_i g_i{}^{N_i} \tag{3.47}$$

Because the particles are distinguishable, an interchange of two particles is another permutation and creates an *additional* microstate for the two particles so interchanged. Thus from the interchange viewpoint, each particle represents one microstate possibility, and the total number of such possibilities evidently is $N!$. But the order in which particles are assigned to a state within a level is not significant; the identities are what matter. For example, if in working the permutations the sequence abc occurs for a particular state, a new microstate is not created by making the sequence bca or cba since the same particles remain in the same state. Therefore we must divide $N!$ by the number of different sequences in which the same particles appear. This number is $N_1!$ for the first level, $N_2!$ for the second level, and so forth to $N_i!$. The total number of ways in which N particles can be interchanged among all the states in all the levels, with N_1 particles in level 1, N_2 particles in level 2, and onward to N_i in level i, is

$$\frac{N!}{N_1!\, N_2!\, \ldots \, N_i!} = \frac{N!}{\prod_i N_i!} \tag{3.48}$$

The total number of microstates in any macrostate j is the product of Eq. (3.47), which accounts for the assignments of particles to states, and Eq.

TABLE 3.2 POSSIBLE MICROSTATES OF TWO DISTINGUISHABLE PARTICLES AND FOUR STATES OF A SINGLE ENERGY LEVEL

State	Possible Microstates (Arrangements)															
	1	2	3	4	5	6	7	8	9	10	11	12	13	14	15	16
g_4	ab				a	a	a	b			b			b		
g_3		ab			b			a	a	a		b			b	
g_2			ab			b			b		a	a	a			b
g_1				ab			b			b			b	a	a	a

(3.48), which accounts for the interchanges of particles. The product is

$$\mathcal{W}_j = \prod_i g_i^{N_i} \times \frac{N!}{\prod_i N_i!} = N! \prod_i \frac{g_i^{N_i}}{N_i!} \tag{3.49}$$

for each level over all levels. This expression should be compared with Eq. (3.43) for Bose–Einstein and with Eq. (3.45) for Fermi–Dirac statistics. Figure 3.15 shows the Boltzmann macrostates for the same system that was employed in the two previous sets of statistics. It is seen that the format and the number of macrostates are the same as in Figure 3.13, because here also we have no restriction on the number of particles per state. The particles are now distinguishable, however, so that we must keep track of each

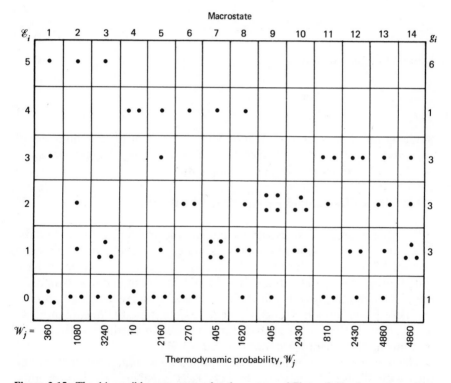

Figure 3.15. The 14 possible macrostates for the system of Figure 3.11 when governed by Boltzmann statistics. After Sears and Salinger (1975). *Introduction to Thermodynamics, Kinetic Theory, and Statistical Thermodynamics,* 3rd ed. © 1975 by Addison-Wesley, Reading, Mass. Used with permission.

particle and where it goes. Therefore, the number of microstates is much greater than in the Bose–Einstein statistics. In Figure 3.15 the thermodynamic probabilities of the macrostates (number of microstates in each) are listed beneath the columns, where the large numbers are noteworthy.

Three calculations of $\mathcal{W}_j$ from Eq. (3.49) and the macrostate arrangements shown in Figure 3.15 are instructive. Taking macrostate 4, we have $N_4 = 2$ and $g_4 = 1$ for level 4, and $N_0 = 3$ and $g_0 = 1$ for level 0. Substitutions into Eq. (3.49) give

$$\mathcal{W}_4 = 5! \left(\frac{1^2}{2!} \times \frac{1^3}{3!} \right) = \frac{5!}{2! \, 3!} = 10$$

Substituting in a similar manner for macrostate 10, we get

$$\mathcal{W}_{10} = 5! \left(\frac{3^3}{3!} \times \frac{3^2}{2!} \right) = 2430$$

Likewise for macrostate 14,

$$\mathcal{W}_{14} = 5! \left(\frac{3^1}{1!} \times \frac{3^1}{1!} \times \frac{3^3}{1!} \right) = 4860$$

Studying the distributions in Figure 3.15 we see that greater numbers of microstates tend to be produced when a level having a great statistical weight is accessible to a large total population.

3.3.3. Bose–Einstein and Fermi–Dirac Distributions

With the statistics of distributions in mind, the problem is now to find the most probable arrangement of particles, states, and levels. As discussed in Section 3.2.5, the length of time that the system spends in any one macrostate is proportional to the number $\mathcal{W}$ of microstates in that macrostate. To this macrostate the system will evolve irreversibly in reaching equilibrium, and here it will remain. Hence we need to determine the maximum of $\mathcal{W}$ in order to find the most probable macrostate. In maximizing $\mathcal{W}$ we are maximizing the entropy according to Eq. (3.41).

The maximum of $\mathcal{W}$ usually is determined by working with $\ln \mathcal{W}$ rather than with $\mathcal{W}$ itself because of the very large numbers. An introductory treatment of the distributions, as here, involves several approximations, but the resulting errors are negligible for large numbers of particles. Assuming that N, N_i, g_i, and also $(g_i - N_i)$ in the Fermi–Dirac case, are large compared to unity, we ignore this factor when it appears in such contexts.

Following generally the plan of Mandl (1980) and Vincenti and Kruger (1965), we carry on a combined development of the two distributions that allows them to be compared at several points.

Taking natural logarithms of Eqs. (3.43) and (3.45) we get

$$\ln \mathcal{W} = \sum_i \left[\ln(N_i + g_i)! - \ln N_i! - \ln g_i!\right] \tag{3.50}$$

for the Bose–Einstein statistics, and

$$\ln \mathcal{W} = \sum_i \left[\ln g_i! - \ln N_i! - \ln(g_i - N_i)\right] \tag{3.51}$$

for the Fermi–Dirac statistics. Employing a simplified version of Stirling's formula for the logarithm of a large factorial number,

$$\ln N! = N \ln N - N$$

allows Eq. (3.50) to be written as

$$\ln \mathcal{W} = \sum_i \left[(N_i + g_i) \ln(N_i + g_i) - N_i \ln N_i - g_i \ln g_i\right] \tag{3.52}$$

and Eq. (3.51) as

$$\ln \mathcal{W} = \sum_i \left[(N_i - g_i) \ln(g_i - N_i) - N_i \ln N_i + g_i \ln g_i\right] \tag{3.53}$$

A word of caution is not amiss. Stirling's formula applies strictly only to very large numbers, but a difference term such as $(g_i - N_i)$ may be small or even zero if the level is fully occupied. Sears and Salinger (1975) may be consulted for a rigorous treatment that avoids Stirling's formula. In the cases of interest here, however, very few of the states are occupied (only one in about 10^7 as shown earlier), so that the condition $g_i \gg N_i$ exists. By regrouping terms we can combine Eqs. (3.52) and (3.53) into

$$\ln \mathcal{W} = \sum_i \pm g_i \ln \frac{g_i \pm N_i}{g_i} + N_i \ln \frac{g_i \pm N_i}{N_i} \tag{3.54}$$

where the plus sign in the $\pm$ refers to Bose–Einstein and the minus sign refers to Fermi–Dirac statistics.

The condition for ln $\mathcal{W}$ of Eq. (3.54) to be a maximum is that small changes δN_i in any of the individual N_i's shall not affect the value of $\ln W_{max}$. Hence we must have

$$\delta \ln W_{max} = \sum_i \left[\pm g_i \ln \frac{g_i \pm N_i}{g_i} + N_i \ln \frac{g_i \pm N_i}{N_i} \right] \delta N = 0 \qquad (3.55)$$

The differentiation is easily performed by operating on Eqs. (3.52) and (3.53) separately and then combining the results, as we did in arriving at Eq. (3.54) originally. Doing this and simplifying, we get

$$\delta \ln W_{max} = \sum_i \left[\ln \frac{g_i \pm N_i}{N_i} \right] \delta Ni = 0 \qquad (3.56)$$

The small changes of δN_i cannot be independent but are related to each other by the constraints (3.38) and (3.39). Since the total number N of particles remains the same for each possible macrostate, any increases in the occupancy numbers N_i of some levels must be balanced by decreases in N_i of other levels. From this reasoning we must have from Eq. (3.38),

$$\sum_i \delta Ni = 0 \qquad (3.57)$$

for the first restriction. In like manner, the total energy $\mathscr{E}$ remains the same for each macrostate. Therefore, any increases in energy $\mathscr{E}_i$ of any level, resulting from increases in N_i, must be balanced by decreases in $\mathscr{E}_i$ of other levels. That is, we must have from Eq. (3.39),

$$\sum_i \mathscr{E}_i \delta N_i = 0 \qquad (3.58)$$

for the second restriction. These two restrictions must be considered together. Equations (3.56), (3.57), and (3.58) are three simultaneous equations that must be satisfied in finding the maximum of $\ln W$.

Equation (3.56) can be solved, subject to the restrictions, by *Lagrange's method of undetermined multipliers*. This involves multiplying Eq. (3.57) by $-\alpha$, and (3.58) by $-\beta$, which gives

$$-\alpha \sum_i \delta N_i = 0$$

$$-\beta \sum_i \mathscr{E}_i \delta N_i = 0$$

Adding these to Eq. (3.56), we have

$$\sum_i \left[\ln\left(\frac{N_i \pm g_i}{N_i}\right) - \alpha - \beta \mathscr{E}_i \right] \delta N_i = 0$$

In effect the δN_i's are now independent of each other. Therefore, the quantity in brackets must vanish for each i, or

$$\ln\left(\frac{N_i \pm g_i}{N_i}\right) - \alpha - \beta \mathscr{E}_i = 0$$

This gives

$$N_i = \frac{g_i}{\exp(\alpha)\,\exp(\beta \mathscr{E}_i) \pm 1} \qquad (3.59)$$

for the values of N_i that make $\ln \mathscr{W}$, and hence $\mathscr{W}$, a maximum in terms of the Lagrangian parameters α and β.

Determining the value of α, actually the term $\exp(\alpha)$, is a straightforward matter in a limiting situation which exists when the particles are thinly distributed over the accessible states. This is true in the atmosphere where most of the states are empty, as shown by Eq. (3.32a) and the discussion there. We saw that only 1 in about 10^7 states is occupied when the gas is at STP conditions. Hence $g_i \gg N_i$ and the only way that Eq. (3.59) can be consistent with this great inequality is for the denominator to be always very large. Since the term $\exp(\beta \varepsilon)$ has a wide range of values, the lowest of which can be unity, this requires that the term $\exp(\alpha)$ be very large. The unity term is then negligible, and we write Eq. (3.59) as

$$N_i = \frac{g_i}{\exp(\alpha)\,\exp(\beta \mathscr{E}_i)} \qquad (3.60)$$

to facilitate matters. Substituting this value of N_i into Eq. (3.38), we get

$$N = \sum_i \frac{g_i}{\exp(\alpha)\,\exp(\beta \mathscr{E}_i)}$$

which introduces the total population N, a new parameter. Solving for $\exp(\alpha)$, which gives the effect of the Lagrangian parameter α, we find

$$\exp(\alpha) = \frac{1}{N} \sum_i g_i \exp(-\beta \mathscr{E}_i)$$

When this is substituted into Eq. (3.59), the result is

$$N_i = N \frac{g_i \exp(-\beta \, \mathscr{E}_i)}{\Sigma_i g_i \exp(-\beta \, \mathscr{E}_i) \pm 1} \tag{3.61}$$

where the plus sign marks the Bose–Einstein and the minus sign the Fermi–Dirac distribution. The value of the parameter β is introduced in the following section.

3.3.4. Boltzmann Distribution

A development of the Boltzmann distribution from the corresponding statistics employs the method used for the other two distributions. Therefore, only the key steps will be shown. Again, we need to find the maximum of $\mathscr{W}_j$, as expressed by the statistics (3.49), in order to determine the most probable macrostate. Taking the logarithm of Eq. (3.49) we have

$$\ln \mathscr{W}_j = \ln N! + \sum_i N_i \ln g_i - \sum_i \ln N! \tag{3.62}$$

Using Stirling's formula

$$\ln N! = N \ln N - N$$

we write for Eq. (3.62),

$$\ln \mathscr{W}_j = N \ln N + \sum_i N_i \ln g_i - \sum_i N_i \ln N_i + \sum_i N_i$$

Since $\Sigma_i N_i = N$ we can regroup terms and have

$$\ln \mathscr{W}_j = N \ln N - \sum_i N_i \ln \frac{g_i}{N_i} \tag{3.63}$$

The maximum of this function is to be found.

Taking the differential of Eq. (3.63) and recognizing that N and g_i are constants, we write

$$\delta \ln \mathscr{W}_j = \sum_i \ln g_i \, \delta \, N_i - \sum_i \ln N_i \, \delta N_i - \sum_i \delta \, N_i \tag{3.64}$$

The logic leading to Eq. (3.57) applies here, as does that expression itself,

$$\sum_i \delta N_i = 0 \tag{3.65}$$

and likewise with Eq. (3.58),

$$\sum_i \mathscr{E}_i \, \delta N_i = 0 \tag{3.66}$$

Since $\ln\mathscr{W}$ is to be a maximum, we must have $\delta \ln\mathscr{W} = 0$. Applying this requirement and Eq. (3.65) to Eq. (3.63), we obtain

$$\sum_i \ln\frac{g_i}{N_i} \, \delta N_i = 0 \tag{3.67}$$

Using Lagrange's method of undetermined multipliers as in the preceding section, we multiply Eq. (3.65) by $-\alpha$ and Eq. (3.66) by $-\beta$, giving

$$-\alpha \sum_i \delta N_i = 0$$

$$-\beta \sum_i \mathscr{E}_i \, \delta N_i = 0$$

Adding these to Eq. (3.67) we get

$$\sum_i \left(\ln\frac{g_i}{N_i} - \alpha - \beta\mathscr{E}_i \right) \delta N_i = 0 \tag{3.68}$$

In effect the δN_i's are now independent, and the coefficient of each one must be zero. Hence for any level i, we have from Eq. (3.68),

$$\ln\frac{g_i}{N_i} - \alpha - \beta\mathscr{E}_i = 0 \tag{3.69}$$

Then

$$N_i = \frac{g_i}{\exp(\alpha) \, \exp(\beta\mathscr{E}_i)} \tag{3.70}$$

Substituting this into Eq. (3.38) we get

$$N = \sum_i \frac{g_i}{\exp(\alpha)\,\exp(\beta\mathscr{E}_i)}$$

which introduces the total population N. Solving for $\exp(\alpha)$ we get

$$\exp(\alpha) = \frac{1}{N} \sum_i g_i \exp(-\beta\mathscr{E}_i) \tag{3.71}$$

When this is substituted into Eq. (3.70), the result is

$$N_i = N\,\frac{g_i\,\exp(-\beta\mathscr{E}_i)}{\Sigma_i\,g_i\,\exp(-\beta\mathscr{E}_i)} \tag{3.72}$$

for the Boltzmann distribution.

The value of the Lagrangian parameter β is derivable from the above expressions and a thermodynamic comparison. Because many manipulations not concerned with the physics are required, the derivation is not given here. Interested readers may consult Beiser (1967), Vincenti and Kruger (1965), or the elegant treatment by Sommerfeld (1964). The derivation yields

$$\beta = \frac{1}{\kappa T} \tag{3.73}$$

Substituting this value into Eqs. (3.61) and (3.72), we have

$$N_i = N\,\frac{g_i\,\exp(\mathscr{E}_i/\kappa T)}{\Sigma_i\,g_i\,\exp(-\mathscr{E}_i/\kappa T) \pm 1} \tag{3.74}$$

for the Bose–Einstein and Fermi–Dirac, and

$$N_i = N\,\frac{g_i\,\exp(-\mathscr{E}_i/\kappa T)}{\Sigma_i\,g_i\,\exp(-\mathscr{E}_i/\kappa T)} \tag{3.75}$$

for the Boltzmann distribution. These are the distributions that produce the maximum numbers of microstates, hence the most probable macrostates, from the three types of statistics.

These expressions differ only by the term ± 1 in the denominator. We recall that only one in about 10^7 energy states is populated when the gas is at STP conditions, so that $g_i \gg N_i$. Advantage was taken of this inequality to eliminate the unity term in the denominator of Eq. (3.59) and arrive at Eq. (3.60). The same factors and reasoning apply here, since the summation term of Eq. (3.60a) appears in the denominator of Eq. (3.74). When this term becomes very large, the unity term in Eq. (3.74) can be neglected. For this condition, known as the *classical* or *Boltzmann limit*, Eq. (3.74) is identical to Eq. (3.75), which henceforth is our concern in dealing with the molecular populations of energy levels.

The maximum calculated previously for ln $\mathcal{W}$, and which has led to Eq. (3.75), is extremely sharp. To demonstrate this, Vincenti and Kruger (1965) consider 1 cm³ of gas at STP conditions and postulate that all values of N_i differ on the average by 10^{-3} from those given by Eq. (3.75). They show that the ratio of the resulting number $\mathcal{W}$ of microstates to the number $\mathcal{W}_{max}$ as determined above is

$$\frac{\mathcal{W}}{\mathcal{W}_{max}} = \exp(-10^{13})$$

This is quite a small number. Hence distributions even slightly different from Eq. (3.75) are so rare as to be negligible. In slightly different words, our knowledge is such that we assign a high degree of probability to the correctness of the Boltzmann distribution, quite in line with the quotation that opens this chapter.

3.3.5. Partition Function, Population Indices

In the preceding sections we have seen the ways in which a total population of particles can be distributed over the energy levels of a system, and we have derived expressions for the most probable distributions. Here we review and summarize several indices for the populations of levels

Statistical weight
Relative population
Partition function
Fractional population
Occupancy number

Although some of these have been mentioned previously in various forms, they are defined here for emphasis and convenience of reference.

Statistical weight expresses the number of states in a given energy level. It is equal to the degeneracy of that level and is designated by the same symbol g. If no other factors were operative, the *a priori* probability of the population ratio of two levels $\mathscr{E}_1$ and $\mathscr{E}_0$ would be

$$\frac{N_1}{N_0} \propto \frac{g_1}{g_0} \qquad (3.76)$$

as in Eq. (3.33). That is, the more states there are within a given level, the greater would be its population. The larger number of states means more rooms, in a probability sense, for particles.

Relative population N_r, as the name implies, expresses the population of a given level in terms of the population of a reference level. Often the reference level is selected to be the lowest of those under consideration. An example will give meaning to this rather empty definition, and for this purpose a simple distribution function is sufficient. We therefore employ the Boltzmann factor, which was singled out from the kinetic energy distribution in Section 2.4.3. This factor,

$$\exp\left(-\frac{\mathscr{E}_i}{\kappa T}\right)$$

gives the occupancy probability of any level i if that level is not degenerate, that is, if $g_i = 1$. Then a function for the population of a level $\mathscr{E}_1$ relative to the level $\mathscr{E}_0$ is

$$N_r = \frac{N_1}{N_0} = \frac{g_1}{g_0} \frac{\exp(-\mathscr{E}_1/\kappa T)}{\exp(-\mathscr{E}_0/\kappa T)}$$

$$= \frac{g_1}{g_0} \exp\left(\frac{\mathscr{E}_1 - \mathscr{E}_0}{\kappa T}\right) \qquad (3.77)$$

Partition function Z is a normalizing factor that, among other things, relates relative population to fractional population. It is invested with considerable meaning in statistical mechanics, but here we point to its role in normalizing a distribution. For this purpose we need only to define it by

$$Z = \sum_i g_i \exp\left(-\frac{\mathscr{E}_i}{\kappa T}\right) \qquad (3.78)$$

The partition function is often called the *sum over states,* although in the terminology of this volume a more appropriate name would be the *sum over levels.* Writing out the first few terms of the summation,

$$Z = g_1 \exp\left(-\frac{\mathscr{E}_1}{\kappa T}\right) + g_2 \exp\left(-\frac{\mathscr{E}_2}{\kappa T}\right) + g_3 \exp\left(-\frac{\mathscr{E}_3}{\kappa T}\right) + \ldots \qquad (3.79)$$

we see that each term is proportional to the number of particles in that level. Thus Eq. (3.79) conveys the essence of the normalizing role in that Z is proportional to the sum of the relative populations of all the levels. Therefore, division of any *relative* population by the partition function gives the *fractional* population of that level. We have met Eq. (3.78) before as the denominator of the Boltzmann distribution, Eq. (3.75).

Fractional population N_f of a level is a simple numeric, where $0 < N_f < 1$, and provides a quick view of a population spread. A normalization requirement

$$\sum_i N_{f,\,i} = 1 \qquad (3.80)$$

obviously holds. From the discussion of the partition function Z we see that

$$N_{f,\,i} = \frac{N_{r,\,i}}{\Sigma_i\,N_{r,\,i}} = \frac{N_{r,\,i}}{Z} = \frac{g_i\,\exp(-\mathscr{E}_i/\kappa T)}{\Sigma_i\,g_i\,\exp(-\mathscr{E}_i/\kappa T)} \qquad (3.81)$$

Summed over all levels, this meets the requirement (3.80).

Occupancy number N_i, or absolute population, of a level is discussed in Section 3.2.2 in connection with Figure 3.9. There it is defined as the sum of the numbers of particles in the states of a given level. In the context of population indices it is the product of the fractional population $N_{f,\,i}$ of the level and the total population N of the system. From the meaning of fractional population Eq. (3.81) we have

$$N_i = N \times N_{f,i} = N\frac{N_{r,\,i}}{Z}$$

$$= N\frac{g_i\,\exp(-\mathscr{E}_i/\kappa T)}{\Sigma_i\,g_i\,\exp(-\mathscr{E}_i/\kappa T)} \qquad (3.82)$$

which is identical to Eq. (3.75).

4

Molecular Internal Energies

This chapter deals with the three forms of molecular internal energy, the associated energy levels, and the quantum rules for transitions between levels. First we review the elements of a molecular dipole moment and the extent to which atmospheric species possess such moments. Rotational energy and transitions are treated next, starting with the concept of a classical rotator into which are introduced the quantum restrictions on angular momentum. Vibrational energy is similarly treated by extending the concept of a classical vibrator. The vibrational modes of diatomic and polyatomic molecules are explained. Electronic energy states and their extensive nomenclature are then reviewed, and the numerous selection rules for transitions are discussed.

4.1. MOLECULAR DIPOLE MOMENTS

Radiative transitions require that the molecule possess an electric or magnetic dipole moment. Interaction of the dipole moment and the electromagnetic field effectively couples the molecule and the field. The extent to which a molecule has a permanent dipole moment, or can acquire an induced one when subjected to a field, relates to the polarizability of the molecule. Upon this polarizability depend the properties of dielectric constant and refractive index, which are exhibited by substances in bulk form. Here we are concerned with the electric dipole moment itself and the molecular species having such moments. Certain species also exhibit magnetic dipole moments, but since the moments are only about 10^{-5} of the electric type they are not discussed here.

A molecule has a dipole moment when its effective centers of positive and negative charge are separated. More specifically the moment $\boldsymbol{\mu}$, which is a vector quantity, is defined by

$$\boldsymbol{\mu} = \sum_i e_i \mathbf{d}_i \tag{4.1}$$

where the summation extends over all nuclear and electronic charges e_i of the molecule. The position vectors $\mathbf{d}_i$ may be referred to any origin when the molecule has no net charge, that is, when the molecule is non-ionized.

Here we consider that the charge distribution over the molecule can be represented by centers of positive and negative charges having value Q and separation $\mathbf{d}$. For this simplification the moment vector $\boldsymbol{\mu}$ has the magnitude

$$\mu = |Q \times \mathbf{d}| \tag{4.2}$$

as illustrated in Figure 4.1. The vector points from the negative to the positive pole, and the electric field of the dipole has cylindrical symmetry.

The separation of the charges (nuclei) constituting the dipole is about 0.10 nm or 10^{-8} cm, and the magnitude of the displaced charge is comparable to the charge e of one electron having the value

$$e = 4.80 \times 10^{-10} \text{ statcoulomb (an esu unit in the CGS system)}$$

$$= 1.60 \times 10^{-19} \text{ coulomb} \quad \text{(the SI unit)} \tag{4.3}$$

Thus early workers found it convenient to express electric dipole moments in the CGS units *statcoulomb centimeter,* named the *debye* after the Dutch physicist. The numerical value is

$$1 \text{ debye} = 10^{-18} \text{ statcoulomb centimeter}$$

$$= 3.33 \times 10^{-30} \text{ coulomb meter}$$

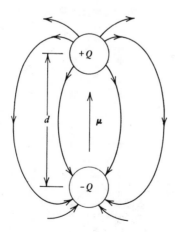

Figure 4.1. Elements of an electric dipole moment. From von Hippel (1954). Used with permission of MIT Press, Massachusetts Institute of Technology.

Much of the current work is still reported in CGS units. The dimensions of the debye correspond to the product of charge and distance, and for CGS units they are $M^{1/2}L^{5/2}T^{-1}$. For SI units they are $M^{1/2}L^{3/2}$.

Configurations of the nuclei in atmospheric molecules are shown in Figure 4.2, and the values of the moments are listed in Table 4.1. The homonuclear diatomic molecules have symmetrical charge distributions, hence no electric dipole moments. Consequently, these gases do not undergo transitions in rotational and vibrational energy and have no radiative activity in the infrared. Both species do have weak magnetic dipole moments that provide some coupling with the electromagnetic field, which enables transitions in electronic energy to occur. The heteronuclear, diatomic CO molecule has an asymmetrical charge distribution, hence a permanent electric moment. The CO_2 molecule is a linear array with a symmetrical charge

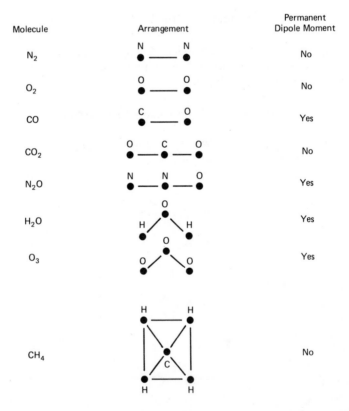

Figure 4.2. Nuclear configurations and permanent dipole moment status of atmospheric molecules.

TABLE 4.1 VALUES OF MOLECULAR ELECTRICAL DIPOLE
MOMENT μ IN DEBYE UNITS

			Molecular Species			
CO	CO_2	N_2O	H_2O	O_3	CH_4	NO_2
0.112	0	0.167	1.85	0.53	0	0.316

Source: Nelson (1967).

distribution, as may be seen in Figure 4.2, hence has no permanent moment. The N_2O molecule is also a linear array, but has an asymmetrical charge distribution, hence a permanent moment. Bent molecules such as H_2O and O_3, known as asymmetric tops, have asymmetric distributions of charge and so have permanent moments. The CH_4 molecule, known as a spherical top, has no permanent moment.

Radiative transitions of purely rotational energy require that the molecule possess a permanent electric dipole moment. Thus CO, N_2O, H_2O, and O_3 exhibit pure rotation spectra, while N_2, O_2, CO_2, and CH_4 do not. Radiative transitions of vibrational energy require a change in dipole moment during the vibration cycle, that is, an oscillating dipole moment. This requirement is met by CO_2 in two of its three vibrational modes, and CH_4 in two of its four modes. An oscillating moment is also effective for rotational transitions, hence CO_2 and CH_4 each have two vibration–rotation bands. The permanent moments of CO, N_2O, H_2O, and O_3 assume oscillating values, of course, during vibration. Thus the first of these species, having only a single vibrational mode, has a single vibration–rotation band. Each of the remaining species has three vibrational modes, hence three vibration–rotation bands. All of these bands, which have varying widths, can be seen in Figure 1.1.

4.2. ROTATIONAL ENERGY AND TRANSITIONS

In this section we consider first a classical rotator from the macroscopic world and its angular momentum and kinetic energy. The concept of reduced mass is described. This rotator is then transformed into a quantized rotating molecule by introducing the quantum restrictions on angular momentum, hence on kinetic energy. Transitions between energy levels, and the resulting spectral lines, are reviewed. Attention is given to the degeneracy of rotational energy levels. In conclusion, the concepts of partition function and population indices are applied to the rotational levels, and expressions for calculating their populations are given.

4.2.1. A Classical Rotator

The total rotation of a rigid body can be defined by the components about three orthogonal axes originating at its center of gravity. The axes of rotational freedom of linear and asymmetric top molecules are indicated in Figure 4.3. For a diatomic or linear triatomic molecule the moment of inertia about the internuclear axis (bond line) is taken to be zero because the nuclei are *point masses*. Consequently, neither type of structure can rotate about this axis. Such molecules have two equal moments of inertia and two degrees of rotational freedom. Asymmetric top molecules, also known as *bent triatomic molecules,* have three unequal moments and three degrees of rotational freedom. Molecular structures of greater complexity have additional freedoms. In all cases the degrees of rotational freedom are additional to the three degrees of translational freedom.

Relationships between the parameters of rotational energy are easily seen for a diatomic molecule. They apply also to the degrees of rotational freedom of polyatomic molecules, with due regard for interactions or coupling among the axes. Taking the diatomic case illustrated in Figure 4.4, we have two nuclear masses m_1 and m_2 at distances r_1 and r_2 from their common

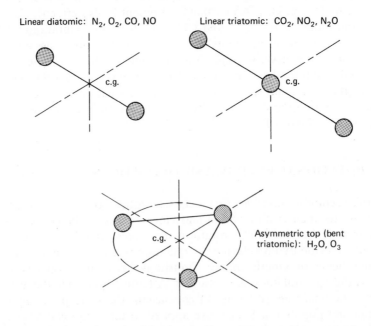

Figure 4.3. Axes of rotational freedom for linear and asymetric top molecules.

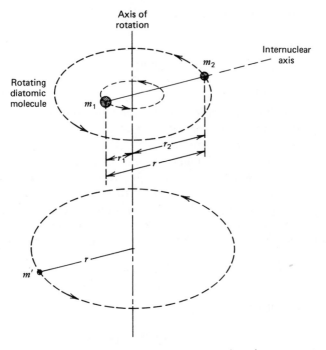

Figure 4.4. Reduced mass of a molecule. From Eisberg and Resnick (1974).

center of gravity. The moment of inertia of this two-mass rigid rotator (dumbbell) is

$$I = m_1 r_1^2 + m_2 r_2^2$$

From this we derive

$$r_1 = \frac{m_2}{m_1 + m_2} r$$

$$r_2 = \frac{m_1}{m_1 + m_2} r \qquad (4.4)$$

where r is the internuclear distance. Substitutions in the initial expression give

$$I = \left(\frac{m_1 m_2}{m_1 + m_2}\right) r^2 = m' r^2 \qquad (4.5)$$

The dimensions of I evidently are ML^2. In Eq. (4.5) m' is the *reduced mass* defined by

$$m' = \frac{m_1 m_2}{m_1 + m_2} \tag{4.6}$$

This concept is helpful in that it permits calculating the moment of inertia without regard for the actual location of the center of gravity.

As an example we find the moment of inertia of the CO molecule from available data. The most abundant isotopes of atmospheric carbon and oxygen have atomic weights of 12 and 16, respectively. Dividing these values by Avogadro's number $N_A = 6.023 \times 10^{-23}$, in the sense of Eq. (2.8), gives 1.99×10^{-23} g and 2.66×10^{-23} g for the atomic masses of ^{12}C and ^{16}O. The internuclear distance C—O is 0.13nm $= 1.13 \times 10^{-8}$ cm, as determined by spectroscopy. Then from Eq. (4.6) the reduced mass has the value

$$m' = 1.14 \times 10^{-23} \text{ g} \tag{4.6a}$$

and the moment of inertia is

$$I = 1.46 \times 10^{-39} \text{ g cm}^2 \tag{4.7}$$

The moment of inertia of any diatomic molecule is calculated in a similar way.

At this stage we assume that all of the molecular species that will be considered as examples are in the electronic ground state. This means that the angular momentum of a molecule is due entirely to rotation of the nuclei about their common center of gravity. The angular momentum of a rigid rotator is defined by

$$\text{A.M.} = I\omega = m'r^2\omega$$

where ω is the angular velocity in rad sec^{-1}. Angular momentum is a vector quantity whose dimensions are ML^2T^{-1}. The vector lies along the rotational axis, and it points in the direction that a right-hand screw would advance if rotated in the same sense as the body. The kinetic energy $\mathcal{E}_{rot}$ of a rotator is equal to one-half the product of angular momentum and angular velocity, or

$$\mathcal{E}_{rot} = \frac{1}{2} I\omega^2 \tag{4.8}$$

The dimensions of rotational kinetic energy are ML^2T^{-2}, the same as those of the translational form. When I is stated in g cm^2 and ω in rad sec^{-1}, the unit of $\mathscr{E}_{rot}$ is the erg ($=10^{-7}$ J). In the macroscopic world, as exemplified by a classical rotator, both angular momentum and rotational kinetic energy have continuous ranges of values.

4.2.2. A Quantized Rotator

Quantum restrictions on rotational energy are a consequence of the quantum restrictions on angular momentum. These latter restrictions are found from a solution of the time-independent Schroedinger equation (3.14). Substituting therein the *reduced mass* m' for m and zero for $\mathscr{E}_p$, because rotation does not involve potential energy if the rotator is rigid, the equation then reads

$$\frac{\partial^2\psi}{\partial x^2} + \frac{\partial^2\psi}{\partial y^2} + \frac{\partial^2\psi}{\partial z^2} + \frac{8\pi^2 m'}{h^2}\mathscr{E}\psi = 0 \tag{4.9}$$

Solutions of this equation are worked out by Pauling and Wilson (1935) and Sommerfeld (1930); we employ the principal result in the following treatment.

The quantum restrictions on angular momentum are given by

$$I\omega = \frac{h}{2\pi}[J(J+1)]^{1/2} \tag{4.10}$$

where J, the *quantum number for rotation,* can assume only the integer values 0, 1, 2, 3. . . . The molecule cannot rotate at angular velocities that would produce momentum values different from those permitted by Eq. (4.10). Quantum restrictions on the rotational energy $\mathscr{E}_{rot}$ now are introduced. Writing Eq. (4.8) as

$$\mathscr{E}_{rot} = \frac{1}{2}\frac{(I\omega)^2}{I} \tag{4.11}$$

and substituting therein the value of $I\omega$ from Eq. (4.10) gives

$$\mathscr{E}_J = \frac{h^2}{8\pi^2 I}J(J+1) \tag{4.12}$$

for the permitted energy values of molecular rotation. When I is in g cm^2 and h is in erg sec, the unit of $\mathscr{E}_J$ is *erg.*

Figure 4.5 shows the energy levels of a CO molecule as calculated from Eq. (4.12) and the value of I in Eq. (4.7). As J increases, the spacing between levels becomes greater because of the J^2 term in Eq. (4.12). However, adjacent differences always differ by

$$2\frac{h^2}{8\pi^2 I}$$

which is equal to about 7.61×10^{-16} erg $= 7.61 \times 10^{-23}$ J for a CO molecule. Other atmospheric species have energy levels similar to those in Figure 4.5, depending on their moments of inertia. Also shown is the mean translational kinetic energy $\mathscr{E}_{tr}$ for an air molecule at 0°C. From Eq. (2.41) this energy, which is defined by $3\kappa T/2$, has the value 5.66×10^{-14} erg. All of

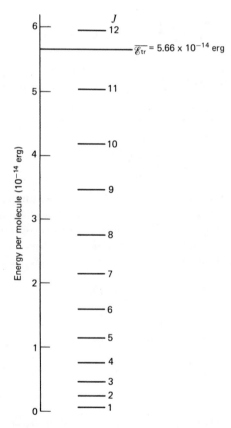

Figure 4.5. Rotational energy levels of a CO molecule as a function of the quantum number.

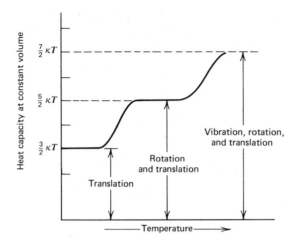

Figure 4.6. Variation of specific heat capacity with temperature, at constant volume, for an idealized gas composed of diatomic molecules.

the atmospheric species have rotational levels whose energy values are less than this. Consequently, when local thermodynamic equilibrium (LTE) exists, many of these levels will be collisionally populated at terrestrial temperatures.

In such a manner the rotational degrees of freedom and associated energy levels provide a means of energy storage in a gas beyond that due to molecular translations. Thus the specific heat, that is, the capacity of a gas for storing thermal energy, increases as gas temperature is raised and more rotational levels become active. This is shown in Figure 4.6 for the energies of both rotation and vibration (anticipating the treatment in Section 4.3.1.). At very low temperatures all of the molecules are in the zero vibrational level and the lowest rotational levels. Energy of the gas is essentially all in the translational form, where each of the three degrees of freedom contributes $\kappa T/2$ to the heat capacity. As temperature increases, more rotational levels become populated and come into action as additional reservoirs for energy. Ultimately, each degree of rotational freedom, of which a diatomic molecule has two, contributes $\kappa T/2$ to the heat capacity, which is then represented by $5\kappa T/2$. At still higher temperatures the vibrational levels above the lowest become active, and the single degree of vibrational freedom of a diatomic molecule contributes $2\kappa T/2$ to the heat capacity, whose total is now $7\kappa T/2$ as in the figure.

Expressions such as Eq. (4.12) can be simplified by grouping the numerical factors and dividing by I to form a *rotational constant*. This is designated B for those species having two equal moments of inertia, such as the diatomic arrangement in Figure 4.3. Species such as asymmetric tops having three moments I_A, I_B, I_C have a rotational constant for each axis, namely, A, B, C. The constant is defined by

$$B = \frac{h}{8\pi^2 c I_B} = \frac{2.80 \times 10^{-39}}{I_B} \tag{4.13}$$

and similarly for the other two. Except for the constant factors in Eq. (4.13) B corresponds to a reciprocal moment of inertia, but its dimension is L^{-1} and the usual unit is cm^{-1}. Table 4.2 lists values of the rotational constants for the common atmospheric species.

In terms of B the energy expression (4.12) becomes

$$\mathcal{E}_J = Bhc \times J(J+1) \tag{4.14}$$

The energy is often expressed by the *rotational term value* $F(J)$, which is defined by

$$F(J) = \frac{\mathcal{E}_J}{hc} = BJ(J+1) \tag{4.15}$$

and measured in the energy unit cm^{-1}.

4.2.3. Rotational Transitions

All of the molecular species shown in Figure 4.2, except N_2 and O_2, meet the electrical dipole requirement for radiative transitions. These transitions obey the quantum selection rule $\Delta J = 1$ for absorption and $\Delta J = -1$ for emission, so that adjacent levels are involved. We now derive expressions for transition energy and the resulting spectral lines. Customarily the upper

TABLE 4.2   ROTATIONAL CONSTANTS OF ATMOSPHERIC MOLECULES

Species	A	B	C
CO	—	1.9314	—
CO_2	—	0.3902	—
N_2O	—	0.4190	—
H_2O	27.877	14.512	9.285
O_3	3.553	0.445	0.395
CH_4	—	5.249	—

Source: Herzberg and Herzberg (1957).

energy level is denoted by $\mathscr{E}'$ and the lower by $\mathscr{E}''$; the quantum number of the upper level is denoted by J' and the number of the lower level by J''. For any two adjacent levels we have from Eqs. (4.14) and (4.15) that the energies are

$$\text{Upper:} \quad \mathscr{E}_J' = BhcJ'(J'+1) \qquad \text{(erg)}$$

$$F(J) = BJ'(J'+1) \qquad \text{(cm}^{-1}) \qquad (4.16a)$$

$$\text{Lower:} \quad \mathscr{E}_J'' = BhcJ''(J''+1) \qquad \text{(erg)}$$

$$F(J'') = BJ''(J''+1) \qquad \text{(cm}^{-1}) \qquad (4.16b)$$

The energy of the transition is equal to the energy difference of the two levels,

$$\Delta\mathscr{E}_J = 2Bhc(J''+1) = 2BhcJ' \qquad \text{(erg)} \qquad (4.17a)$$

$$\Delta F(J) = 2B(J''+1) = 2BJ' \qquad \text{(cm}^{-1}) \qquad (4.17b)$$

Here J' takes the values 1, 2, 3, ... and J'' takes the corresponding values 0, 1, 2, ... according to the rule $\Delta J = \pm 1$.

The absorbed or emitted photons must have energies that are individually equal to Eq. (4.17), and the aggregate effect from all the molecules is observed as a spectral line. The parameters of the line are found by substituting Eq. (4.17a) for $\mathscr{E}$ in Eq. (1.14) to give

$$\nu = 2BcJ' \qquad \text{(Hz)} \qquad (4.18a)$$

$$\lambda = \frac{1}{2BJ'} \qquad \text{(cm)} \qquad (4.18b)$$

$$\bar{\nu} = 2BJ' \qquad \text{(cm}^{-1}) \qquad (4.18c)$$

where J' may have any of the quantum number values 1, 2, 3. ... . Because $\Delta J = \pm 1$, it follows from Eq. (4.18c) that the separation in wavenumber units of adjacent lines is just $2B$ cm^{-1}. Also we see from Eq. (4.16) that the difference of term values for the upper and lower levels of the transition give the wavenumber of the line directly.

We note broadly the use of rotational data to determine several factors of molecular structure. For a diatomic molecule, a measurement of the separations of spectral lines in terms of B allows the moment of inertia to be calculated, as from Eq. (4.13). Masses of the individual atoms are known accurately from chemistry, which enables the reduced mass to be found from Eq. (4.5). From this the internuclear distance or bond length can be calculated from Eq. (4.4). For an asymmetric top molecule, such as H_2O in Figure 4.3, where the bond lengths are determined by other means, a knowledge of I allows the bond angle to be calculated.

Pure rotational lines lie in the extreme infrared and millimeter regions, as may be seen by substituting values of B from Table 4.2 into any of the expressions (4.18). Because J can assume only integer values, the lines nominally are equally spaced. At higher values of J, that is, at higher rotational speeds, centrifugal stretching of the valence bond allows the internuclear distance to increase slightly, which produces larger values of I. This reduces the value of B, and the spectral results may be appreciated from Eq. (4.18). This effect, known as *centrifugal stretching*, is accounted for in the full theory by introducing correction terms containing higher orders of J. Additional terms may be employed to compensate a slight variation of the moment of inertia with vibrational amplitude.

Measurements of rotational factors in microwave spectroscopy are made very accurately by tuned circuits in terms of frequency. Precise determinations of line spacing, hence of the constant B, are achieved at quite low values of J where centrifugal stretching is a minimum. The constant B is defined by

$$B \text{ (microwave)} = \frac{h}{8\pi^2 I}$$

which has the dimension T^{-1}, that of frequency. The methods are such that the value of B comes out in Hz, and this allows an interesting correlation. We saw previously that B in optical spectroscopy is defined by Eq. (4.13), which is repeated here

$$B \text{ (optical)} = \frac{h}{8\pi^2 c I}$$

and has the dimension L^{-1}. Hence the ratio of the microwave to the optical definition has the dimensions LT^{-1}, those of velocity. The value of the ratio is just the term c, which is the speed of light. This is easy to verify numerically. Townes and Schalow (1955) give the value $B = 5.7898 \times 10^{10}$ Hz for the CO molecule. Using the corresponding value of B in Table 4.2, we have the ratio

$$\frac{B \text{ (microwave)}}{B \text{ (optical)}} = \frac{5.7898 \times 10^{10}}{1.9314 \text{ cm}^{-1}} = 2.9979 \times 10^{10} \text{ cm sec}^{-1}$$

which compares well with the value in Appendix A. Plyler et al. (1955) have

accurately measured the molecular constants of CO for such determinations, which are unique in thus employing data from two well-
separated spectral regions.

Returning to the optical region, the molecular rotation rate at a given
energy level is readily calculated. Using J' to denote the quantum number,
Eq. (4.12) becomes

$$\mathscr{E}_{J'} = \frac{h^2}{8\pi^2 I} J'(J' + 1)$$

Substituting for $\mathscr{E}_{J'}$ the classical value of rotational energy from Eq. (4.8)
we get

$$\omega = \frac{h}{2\pi I}[J'(J' + 1)]^{1/2} \tag{4.19}$$

for the angular velocity in rad $\sec^{-1}$. We realize that

$$\omega = 2\pi \nu_{rot}$$

and

$$[J'(J' + 1)]^{1/2} \approx J'$$

Substituting these into Eq. (4.19) we get for the rotational frequency in rev
$\sec^{-1}$,

$$\nu_{rot} \approx \frac{h}{4\pi^2 I} J'$$

$$\approx 2BcJ' \tag{4.20}$$

where B is the rotational constant defined by Eq. (4.13). A comparison of
Eqs. (4.20) and (4.18a) shows that the rotational frequency, at a given
energy level $\mathscr{E}_{J'}$, is approximately equal to the *spectral* frequency of the
transition that has this level $\mathscr{E}_{J'}$ as the upper level. For an example, the
rotational frequency of the CO molecule at the level $J' = 7$ is about 8×10^{11}
rev $\sec^{-1}$.

4.2.4. Populations of Energy Levels

According to the principles of LTE, rotational energy levels acquire their
populations through the medium of molecular collisions, and the populations follow the Boltzmann distribution Eq. (3.75). Because spectral line

strength depends on the number of molecules in the initial level of the transition, we examine the way in which this distribution spreads the population over the levels. These levels are degenerate, each one consisting of multiple states having equal energies. In determining the populations, we first look at the degeneracy g_i and the partition function z.

Existence of the degeneracy can be disclosed conceptually by subjecting a rotating molecule to an electric field, as indicated in Figure 4.7. We denote by $\mathbf{J}$ the vector for angular momentum of the rotation; this symbol is consistent with the notation employed later for electronic states. This vector is parallel to the rotational axis of the molecule. From Eq. (4.10) the magnitude of $\mathbf{J}$ is

$$\mathbf{J} = \frac{h}{2\pi}[J(J+1)]^{1/2} \tag{4.21}$$

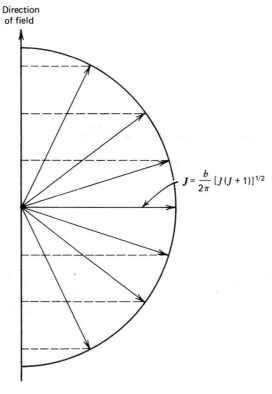

Figure 4.7. Space quantization of rotational angular momentum. Value of the vector component in the direction of the field is always an integer multiple of $h/2\pi$. The figure is drawn for $J = 3$.

The electric field and the rotating dipole of the molecule interact so that **J** never lines up with the direction of the field, but precesses about it in the manner of a gyroscope axis. The resulting angle between **J** and the field direction cannot be arbitrary, and this is the crucial fact. The molecule orients its rotational axis so that the component of **J** in the field direction is always some integer multiple of the quantity $h/2\pi$. This is *spatial quantization,* first demonstrated by Stern and Gerlach (1922).

In Figure 4.7 the quantized component in the field direction remains constant in magnitude for a given value of J, while vector **J** traces out a cone in space with the field direction as an axis. Because of this precessional motion, **J** cannot align itself with the field. Here is a situation where the maximum component of a vector can never be as great as the vector itself. The component can be either parallel (in the *same* direction) with the field or antiparallel (in the *opposite* direction). It is seen from the figure, drawn for $J = 3$, that for any Jth energy level the molecule has $(2J + 1)$ possible orientations, hence it has this number of energy states. All of these states have equal probabilities of occupancy. They are *explicit* in the presence of a field and constitute the degeneracy of that energy level.

Next, let the field gradually be reduced to zero. At no point will any of these states disappear. Instead, they will coalesce and become *implicit* in the one energy level whose value is defined by Eq. (4.12) or (4.14). The multiplicity is now latent, but the degeneracy remains. The existence of these states affects the Boltzmann distribution because each one has an equal chance of occupancy by a molecule at that energy level. Thus the degeneracy factor $(2J + 1)$ is the statistical weight that must be applied to the distribution. It means that if there are $(2J + 1)$ states in a given energy level, there will be $(2J + 1)$ times as many molecules in that level than if there were but a single state. Thus the term $(2J + 1)$ performs the same function as the term g_i used to designate the statistical weight in Section 3.3.5.

The *non-normalized* probability that a given nondegenerate level is occupied is proportional to the Boltzmann factor $\exp\,(-\kappa T)$ discussed in Sections 2.4.3 and 3.3.5. Introducing the statistical weight $(2J + 1)$ and the occupancy number N_J, we express the occupancy probability by

$$N_J \approx (2J + 1)\,\exp\left(-\frac{\mathscr{E}}{\kappa T}\right)$$

Substituting for $\mathscr{E}$ its value from Eq. (4.14) we have

$$N_J \approx (2J + 1)\,\exp\left[-\frac{BhcJ(J + 1)}{\kappa T}\right]$$

which is still only a proportional relationship because the probability is not normalized. Normalization is provided by the partition function discussed below.

Next we consider the relative population N_r, defined generally by Eq. (3.77), and employ the level $J = 0$ as a convenient datum. At this value of J the statistical weight of that level is unity, and the entire exponential term of the above expression becomes unity. For these conditions we see from Eq. (3.77) that the population of any Jth level, *relative* to the population at $J = 0$, is

$$N_r = \frac{N_J}{N_0} = \frac{N_J}{1} = (2J + 1) \exp\left[-\frac{BhcJ(J + 1)}{\kappa T}\right] \tag{4.22}$$

The right hand side of this expression is still the ratio N_J/N_0.

In Eq. (4.22) N_{J_r} is the relative population of the Jth level; N_0 is the population of the level $J = 0$; and the numerator of the exponential is Eq. (4.14). The denominator represents the share of energy that a molecule can claim as a function of temperature, according to Eq. (2.45). Relative populations of the energy levels of the CO molecule previously arrayed in Figure 4.5 are shown in Figure 4.8, calculated from Eq. (4.22). The length of each bar is proportional to N_r for that level, with respect to N_0 normalized to unity. The level containing the maximum population is found by setting the derivative of Eq. (4.22) equal to zero and solving for J. This gives

$$J_{max} = \left(\frac{\kappa T}{2Bhc}\right)^{1/2} - 0.5 = (0.59)^{1/2}\frac{T}{B} - 0.5 \tag{4.23}$$

which does not yield an integer value because Eq. (4.22) has been treated as a continuous distribution. However, the nearest integer value to the solution of Eq. (4.23) is that of J_{max}. For example, a solution for the CO molecule at 300 K gives the value 6.85, or $J_{max} = 7$, in agreement with Figure 4.8.

In order to find the fractional and absolute populations of levels, we must deal with the rotational partition function Z, generally defined by Eq. (3.78), and the total population N. The partition function for rotational energy is defined here by a summation of Eq. (4.22) over all levels,

$$Z = \sum_i (2J + 1) \exp\left[-\frac{BhcJ(J + 1)}{\kappa T}\right] \tag{4.24}$$

Written out for the first three values of J this becomes

$$Z = 1 + 3 \exp\left(-\frac{2Bhc}{\kappa T}\right) + 5 \exp\left(-\frac{6Bhc}{\kappa T}\right) + \cdots \tag{4.25}$$

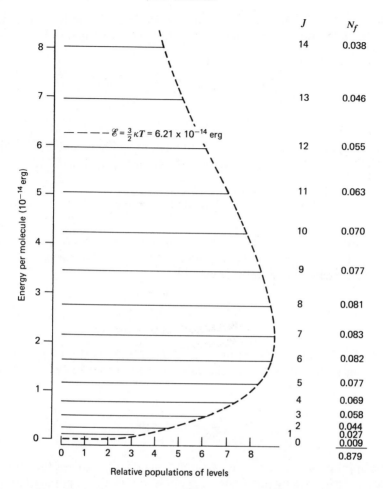

Figure 4.8. Relative populations of rotational energy levels, with degeneracy of $2J + 1$, for CO molecule at $300\ K$. The value $\mathscr{E} = 6.21 \times 10^{-14}$ erg is the translational energy at $300\ K$.

For large values of T or small values of B, Eq. (4.25) can be expressed as an integral,

$$Z = \int_0^\infty (2J + 1) \exp\left[-\frac{BhcJ(J + 1)}{\kappa T}\right]$$

$$= \frac{\kappa T}{Bhc} \tag{4.26}$$

The fractional population N_f is found by dividing the relative population function (4.22) by the partition function (4.26) in the manner of Eq. (3.81). This gives

$$N_f = \frac{Bhc}{\kappa T}(2J + 1) \exp\left[-\frac{BhcJ(J + 1)}{\kappa T}\right] \qquad (4.27)$$

which must meet the normalization requirement (3.80). The occupancy numbers N_J of rotational levels are obtained by multiplying the fractional values (4.27) by the total population N, according to Eq. (3.82). Thus,

$$N_J = N\frac{Bhc}{\kappa T}(2J + 1) \exp\left[-\frac{BhcJ(J + 1)}{\kappa T}\right] \qquad (4.28)$$

It is instructive to derive the fractional from the relative populations shown in Figure 4.8 as calculated from Eq. (4.22). Looking at the function (4.26) and excluding the term B we find

$$\frac{\kappa T}{hc} = 2.08 \times 10^2 \qquad (4.29)$$

for $T = 300$ K. Multiplication by a factor $T/300$ adjusts Eq. (4.29) for any other temperature. Division by the value $B = 1.931$ for CO from Table 4.2 gives

$$Z = \frac{\kappa T}{Bhc} = 1.08 \times 10^2$$

as the appropriate value of the partition function. Division of the abscissa values of Figure 4.8 by this value of Z yields the fractional populations N_f, which essentially are probabilities whose sum for all levels is unity. These values of N_f are shown in the column at the right of the figure, where it is seen that the sum is 0.879, but not unity. The difference is due entirely to ignoring the levels above $J = 14$.

4.3.   VIBRATIONAL ENERGY AND TRANSITIONS

Vibrational energy is treated in this section by first reviewing the parameters of a classical vibrator whose amplitudes and energies have continuous ranges of values. The quantum restrictions from the Schroedinger equations are then introduced, and the quantized nature of the vibrator is examined. An application of the uncertainty principle to the vibrator is made, and the effects of anharmonic vibrations are noted. Energy levels, populations of levels, and transitions are described. The section concludes with a discussion of the vibrational modes available to polyatomic molecules.

4.3.1. A Classical Vibrator

A molecular assemblage of nuclear masses held together by elastic valence bonds forms a system capable of vibrating in one or more modes. The natural (resonant) frequency of this vibrator depends on the nuclear masses and the force constant of the elastic bond, as with a macroscopic vibrator. The number and arrangement of masses and binding forces determine the number of vibrational modes. Diatomic molecules have one mode while triatomic molecules have three. The value of the force constant usually is affected by the electronic energy state. At this point we premise that (1) the molecule remains in the electronic ground state and (2) the rotational quantum number remains at sufficiently low values that centrifugal stretching of the valence bond does not occur. That is, we regard the molecule as a rigid rotator even though it is vibrating.

The vibrational model of a diatomic molecule as shown in Figure 4.9a consists of masses m_1 and m_2 connected by a spring whose force constant is k. At small excursions, within the range of Hooke's law, the value of k does not change as the masses vibrate. The restoring force thus remains linear with displacement. From Newton's second law of motion the behavior of the two masses is described by

$$-k(r - r_0) = m_1 \frac{d^2 r_1}{dt^2}$$

$$-k(r - r_0) = m_2 \frac{d^2 r_2}{dt^2} \qquad (4.30)$$

Here r_1 and r_2 are the distances of the two nuclear masses from their common center of gravity, r is the actual internuclear distance, and r_0 is the equilibrium distance. These designations are employed also in Figure 4.4. If the two masses are unequal, the amplitudes of vibration are different, but the motions are still simple harmonic and of the same fundamental frequency. At larger amplitudes, however, the restoring force tends to become nonlinear with displacement. The vibrator then generates harmonic frequencies (overtones) in addition to the fundamental.

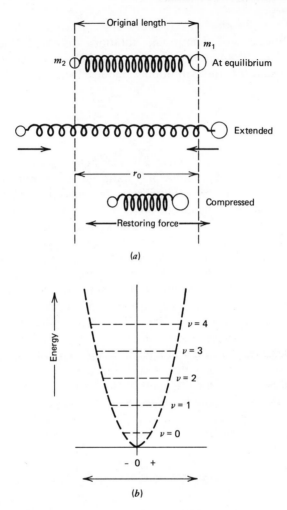

Figure 4.9. (*a*) Two-mass and spring vibrating system and (*b*) potential energy curve with quantized levels. After Barrow (1964). *The Structure of Molecules.* © 1963 by Benjamin/Cummings Publishing Co., Menlo Park, California.

By employing the relationship (4.4) we get from Eq. (4.30) two identical expressions, either of which is

$$\frac{m_1 m_2}{m_1 + m_2} \frac{d^2 r}{dt^2} = - k(r - r_0)$$

Replacing r under the differential sign by $(r - r_0)$, which is allowable since r_0 is a constant, and employing the reduced mass m', we obtain

$$-k(r - r_0) = m' \frac{d^2(r - r_0)}{dt^2} \tag{4.30a}$$

This means that we can regard the two-mass vibrator as equivalent to a single-mass vibrator whose amplitude is equal to the change of internuclear distance. Total energy of the vibrator at any instant is the sum of its potential and kinetic forms, with a smooth exchange between the two during each vibrational cycle. At maximum excursions, that is, at the velocity reversal points, the energy is all potential; at equilibrium positions where the velocities are a maximum, the energy is all kinetic. At the reversal points the energy is given by

$$\mathscr{E}_p = \frac{1}{2} k (r - r_0)^2 \tag{4.31}$$

This vibrational energy, deriving from an alternately stretched and compressed valence bond, is related to the potential energy of electron arrangement discussed in Section 4.4.1.

The natural frequency of the two-mass vibrator is found by solving Eq. (4.30a), which yields

$$\nu_{\text{vib}} = \frac{1}{2\pi} \left(\frac{k}{m'} \right)^{1/2} \tag{4.32}$$

The force constant thus can be written as

$$k = 4\pi^2 m' \nu_{\text{vib}}^2$$

and we write for the potential energy,

$$\mathscr{E}_p = 2\pi^2 m' (r - r_0)^2 \nu_{\text{vib}}^2$$

In words, the energy of a two-mass vibrator varies as the reduced mass, as the square of the amplitude, and as the frequency squared.

The variation of potential energy through a half-cycle of vibration, as the internuclear distance varies from the equilibrium value, is shown by the parabolic curve of Figure 4.9b. In contrast to the classical rotator which can assume any rotational frequency, the classical vibrator at small amplitudes is constrained to the single frequency defined by Eq. (4.32) and to integer multiples thereof. The frequency of a resonant vibrator thus can be regarded as quantized; this characteristic was noted previously in the discussion of Figure 3.1. The lowest- or zero-energy state corresponds to the system at rest, with the masses remaining at their equilibrium distance from each other. In the molecular world such a zero-energy state would conflict with the uncertainty principle in that the positions and momenta of the masses are then known exactly. This and other difficulties of explaining absorption–emission in terms of a classical vibrator are obviated by the concept of a quantized vibrator.

4.3.2.   A Quantized Vibrator

Quantum restrictions on vibrational energy are found from solutions of the time-independent Schroedinger equation (3.12) in one dimension. Substituting therein the reduced mass m' for m, and the value of vibrator potential energy Eq. (4.31) for $\mathscr{E}_p$, we have

$$\frac{d^2\psi}{dx^2} + \frac{8\pi^2 m'}{h^2}\left(\mathscr{E} - \frac{1}{2}kx^2\right)\psi = 0 \qquad (4.33)$$

Solutions will be found in Pauling and Wilson (1935), Beiser (1967), and Eisberg and Resnick (1974) in detail. The eigenvalues defining the allowed energy levels are

$$\mathscr{E}_v = \left(v + \frac{1}{2}\right)h\nu_{\text{vib}}, \qquad v = 0, 1, 2, \ldots \qquad (4.34)$$

where v is the quantum number for vibration. The energy, expressed as a *term value* $G(v)$, and measured in cm^{-1}, is found by dividing $\mathscr{E}_v$ by hc to give

$$G(v) = \frac{\mathscr{E}_v}{hc} = \left(v + \frac{1}{2}\right)\frac{\nu_{\text{vib}}}{c} = \left(v + \frac{1}{2}\right)\omega' \qquad (4.35)$$

where ω' is the *vibrational frequency* in cm^{-1}.

The energy levels in terms of force constant and reduced mass are obtained by substituting in Eq. (4.34) the value of ν_{vib} from Eq. (4.32), which gives

$$\mathscr{E}_v = \left(v + \frac{1}{2}\right)\frac{h}{2\pi}\left(\frac{k}{m'}\right)^{1/2} \qquad (4.36)$$

According to Eqs. (4.34) through (4.36) the motion, although quantized to discrete values of energy, is still simple harmonic. The energy levels are separated by equal increments, as shown on the curve of Figure 4.9b for the simple vibrator. Amplitude of the motion at a particular value of v is defined by the intersections of the parabola and the corresponding energy level. These are the velocity reversal points, in whose vicinity the nuclear masses classically should spend most of their time. We note that the drawing indicates a sharply limited vibration amplitude at each level; this still-classical view must be modified when the nature of the eigenfunctions is examined.

When the quantum number is small, $v = 0$ or 1, the energy and hence the vibration amplitude are likewise small. The vibration then can be regarded as indeed simple harmonic, well approximated by the parabolic curve of Figure 4.9b. Such a curve, however, is predicated upon linearity of the restoring force, that is, the force constant k must be truly *constant*. This matter is discussed below. Because of the factor $\frac{1}{2}$ in the energy equations such as (4.34) molecules retain some vibrational energy even at the level $v = 0$, as indicated by the placement of that level in the figure. This residual energy, known as *zero-point* energy, apparently persists down to zero Kelvin or nearly so. Thus the system never comes to complete vibrational rest, and a conflict with the uncertainty principle is avoided.

The eigenfunctions ψ_v of the Schroedinger equation (4.33) are Hermite orthogonal functions,

$$\psi_v = N_v \exp\left(-\frac{1}{2}\alpha x^2\right) H_v(\sqrt{\alpha}x) \tag{4.37}$$

Here N_v is a normalizing constant, α represents

$$\alpha = 4\pi^2 m' \nu_{vib}/h = 2\pi(m'k)^{1/2}/h \tag{4.38}$$

and $H_v(\sqrt{\alpha}x)$ is a Hermite polynomial of the vth degree. Solutions of Eq. (4.37) are given by Beiser (1967), Eisberg and Resnick (1974), Herzberg (1961), and Walker and Straw (1967). Values of ψ_v for the first few values of v are plotted in Figure 4.10 along with the resulting probability function ψ_v^2. The classical velocity reversal points noted in Figure 4.9b are indicated here by short vertical lines. It is seen that ψ_v^2 not only has large values in the vicinity of the reversal points but also nonzero values in the regions *beyond* these points. Such regions can never be reached by a classical vibrator. The curves are similar to those in Figure 3.6 for the particle in a shallow well of potential and here we may regard the sides as "soft," permitting the possi-

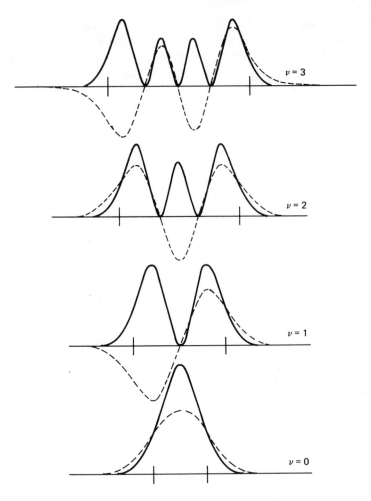

Figure 4.10. Wave functions (dotted lines) and probability densities (solid lines) for a harmonic oscillator at several vibrational levels. From Herzberg (1961). *Molecular Spectra and Molecular Structure, Vol. I. Spectra of Diatomic Molecules.* Copyright © 1950 by Van Nostrand Reinhold Company. Reprinted by permission of Van Nostrand Reinhold Co.

bility of extended excursions. Also, at higher values of v we see that ψ_v^2 has other maxima, in addition to the outermost two, alternating with definite minima.

Behavior of the function at $v = 0$ should be noted. Here there is only one broad maximum at the center instead of one at each reversal region. This central spread of probability, centered about the point of zero energy for a classical vibrator, shows that the eigenfunction is consistent with the uncertainty principle. To carry the idea a little further, Figure 4.11 shows the probability density of the momentum of the oscillating mass for the level $v = 0$, where a fair range of momentum values is seen to be possible. We gather from these two figures that in wave mechanics neither the velocity reversal points nor the maximum momentum (velocity) of the vibrator can be defined exactly. Hence the actual period of vibration is somewhat uncertain, and so is the frequency. The quantity ν_{vib} that we have been using should be regarded as a nominal, or a kind of average, vibrational frequency.

Use of a parabolic curve, such as the one in Figure 4.9b, to represent the potential energy is a simplification that is misleading. That curve predicts an indefinite increase of potential energy as the internuclear distance is increased indefinitely from the equilibrium value. We realize that the restoring force actually becomes zero when the nuclei are separated by a great distance, as when the molecule dissociates. This means that the "force constant" of the valence bond is not really constant, but decreases as the internuclear distance increases. Hence at the larger amplitudes associated with higher values of v the excursions become asymmetrical, particularly in the positive $r - r_0$ direction. The vibration now is increasingly anharmonic, with the result that the frequency is no longer a single fundamental but is marked by several harmonics or overtones of small amplitude.

A more realistic representation of the potential energy is given by the *Morse function*, discussed by Herzberg (1961, 1971), Walker and Straw (1967), and Whiffen (1966) briefly. This function for the H_2 molecule is

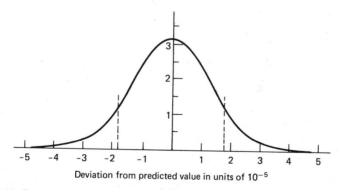

Deviation from predicted value in units of 10^{-5}

Figure 4.11. Probability density of momentum at the zero vibrational level. From Herzberg (1961). *Molecular Spectra and Molecular Structure, Vol. I. Spectra of Diatomic Molecules.* Copyright © 1950 by Van Nostrand Reinhold Company. Reprinted by permission of Van Nostrand Reinhold Co.

illustrated by Figure 4.12, where the strong anharmonicity is quite evident. Several complicating effects follow. Besides the introduction of overtones noted above, the spacing between energy levels decreases as v increases, so that the overtones are less than exact integer multiples of the fundamental. Also the selection rule $\Delta v = \pm 1$ (to anticipate the following subsection) is not always obeyed and transitions occur between nonadjacent levels. This creates overtone spectral frequencies that are relatively weak. Also, we see that at higher levels the average internuclear distance becomes notably greater than the equilibrium distance. Since vibrational frequencies are 10–1000 greater than those of rotation, it is this average distance that determines the moment of inertia. Thus the moment of inertia becomes greater, the rotational constant becomes less, and the frequency spacing of rotational lines decreases, as may be appreciated from Eqs. (4.13) and (4.18).

Finally, Figure 4.12 shows that at the highest levels the restoring force goes to zero. This is the level or point of *dissociation* where the nuclei separate completely. Near and at this region the discrete energy levels of

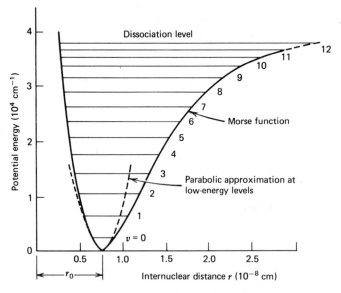

Figure 4.12. Morse curve of vibrational energy for the H_2 molecule in the electronic ground state. The dimension r_0 is the internuclear equilibrium distance. From Herzberg (1961). *Molecular Spectra and Molecular Structure, Vol. I. Spectra of Diatomic Molecules.* Copyright © 1950 by Van Nostrand Reinhold Company. Reprinted by permission of Van Nostrand Reinhold Co.

vibration and rotation are merged into an energy continuum, and spectral lines become broadened into a corresponding continuum. At the low end of the energy scale it is seen that the parabolic approximation is really good only at the level $v = 0$ and possibly $v = 1$. At all higher levels the anharmonicity effects are increasingly important. The foregoing considerations, although discussed here in terms of diatomic molecules, also apply to the several vibrational modes of polyatomic molecules.

4.3.3. Energy Levels, Populations, Transitions

Values of the vibrational energy levels are determined by the eigenvalues (4.34). The amplitude of vibration increases when a transition to a higher level occurs, and decreases for a transition to a lower level. The transition may be either *collisional* by the principles of LTE, or *radiative* by absorption or emission of a photon. A radiative transition requires that the dipole moment change value during the vibration cycle. This requirement naturally is met by a permanent dipole moment.

At this point we consider simple transitions for a diatomic molecule where the initial and final vibrational levels are in the electronic ground state. The quantum selection rules then are $\Delta v = 1$ for absorption and $\Delta v = -1$ for emission. Thus two adjacent levels are involved in either case. Denoting by v' any level except $v = 0$, and by v'' the next lower level, an absorptive transition is indicated by $v' \leftarrow v''$ and an emissive one by $v' \rightarrow v''$. From Eq. (4.36) the transition energy is

$$\Delta \mathscr{E} = \frac{h}{2\pi} \left(\frac{k}{m'} \right)^{1/2} \left[\left(v' + \frac{1}{2} \right) - \left(v'' + \frac{1}{2} \right) \right]$$

$$= \frac{h}{2\pi} \left(\frac{k}{m'} \right)^{1/2} \tag{4.39}$$

because $\Delta v = \pm 1$.

The spectral frequency associated with the transition is found by substituting from Eq. (4.39) the value of $\Delta \mathscr{E}$ for $\mathscr{E}$ in Eq. (1.14). We get

$$\nu = \frac{1}{2\pi} \left(\frac{k}{m'} \right)^{1/2} = \nu_{\text{vib}} \tag{4.40}$$

But from Eq. (4.32) this spectral frequency is also the natural frequency of the vibrator, which is indicated by the last term of Eq. (4.40). Thus the spectral frequency is the same as the vibrator frequency, which does not change when a transition occurs. Only the amplitude changes, producing an energy change according to the term $(r - r_0)^2$ in Eq. (4.31). Evidently a transition between any two adjacent levels should always produce the same spectral frequency. This simple view is satisfactory for the lowest levels, but must be modified for higher levels because of the anharmonicity mentioned earlier.

Energy level and population data are readily found for diatomic species. Considering the CO molecule, its reduced mass from Eq. (4.6a) is $m' = 1.14 \times 10^{-23}$ g. The force constant of the C—O bond has been determined by spectroscopy to be $k = 1.84 \times 10^6$ dyn cm^{-1}. Substituting these values into Eq. (4.36) yields the energy values shown in Figure 4.13, which are notably greater than those in Figure 4.5 for rotational levels. Also shown is the mean value of translational energy $3\kappa T/2 = 6.21 \times 10^{-14}$ erg at 300 K.

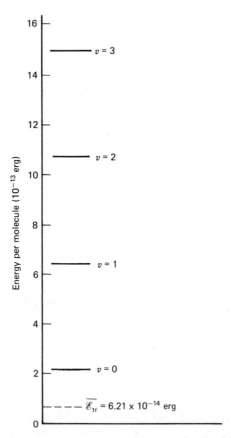

Figure 4.13. Vibrational energy levels of a CO molecule as function of the quantum number.

Since this is less than that of the lowest level of vibration, we expect that relatively few molecules will be collisionally excited to levels above $v = 0$. Because the levels are not degenerate, the statistical weight g is unity and the simple Boltzman distribution (2.61) applies. Substituting therein the following values from Eq. (4.34),

$$\mathscr{E}_{v,1} = 6.42 \times ^{-13} \text{ erg}$$

$$\mathscr{E}_{v,0} = 2.14 \times 10^{-13} \text{ erg} \tag{4.41}$$

and assuming $T = 300$ K, we obtain

$$\frac{N_{v,1}}{N_{v,0}} = 3.261 \times 10^{-5} \tag{4.42}$$

for the population of the level $v = 1$ relative to that of $v = 0$. Similar ratios hold for other atmospheric species at usual terrestrial temperatures.

Transition energy of the CO molecule is found by substituting the previously stated values of m' and k into Eq. (4.39). This gives $\Delta\mathscr{E} = 4.24 \times 10^{-13}$ erg, which agrees well with the difference of the two values in Eq. (4.41). Substituting this value of $\Delta\mathscr{E}$ in $\mathscr{E} = h\nu$ gives the spectral frequency $\nu = 6.43 \times 10^{13}$ Hz. This is equivalent to $\lambda = 4.67$ μm and $\bar{\nu} = 2143$ cm^{-1}, as listed in Table 4.3. Thus the absorption or emission line lies in the middle infrared. We note the value of transition energy in units of cm^{-1}. Multiplication of the $\Delta\mathscr{E}$ value stated above by the conversion factor erg to cm^{-1} listed in Appendix E gives $\bar{\nu} = 2143$ cm^{-1}, which is the wavenumber of the spectral line itself. This relation, that the wavenumber of a spectral line is numerically equal to the energy in cm^{-1} of the associated transition, is important to remember.

4.3.4. Modes of Vibration

Thus far we have considered diatomic molecules because of their simplicity, dealing with CO for convenient examples. The principal radiative species in the atmosphere, however, are polyatomic molecules such as H_2O, CO_2, and O_3. To these we could add SO_2, NO_2, CH_4, and the complex halogen molecules associated with aerosol dispensation of compounds ranging from oven cleaners to antiperspirants. Increasingly, the halogens are viewed as a potential threat to the ozone layer. All polyatomic molecules have *vibrational modes,* which are functions of the number and configuration of nuclear masses, elastic bonds, and interactions. The principal species we should study are represented in Figure 4.14.

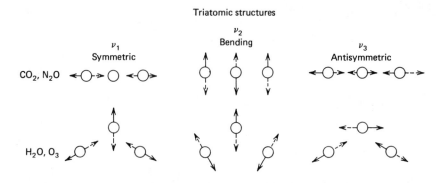

Figure 4.14. Configurations and vibrational modes of diatomic and triatomic atmospheric molecules.

Considering first the diatomic structure, its two nuclei can only move toward and away from each other during vibration. Hence diatomic molecules have but one vibrational mode known as symmetric stretch and designated ν_1. The permanent dipole moment (see Figure 4.1) of the CO molecule assumes an oscillating value during symmetric stretch. The N_2 and O_2 molecules, lacking permanent dipole moments because of symmetrical charge distributions, clearly cannot acquire oscillating moments during this symmetrical vibration. As a result these species have but little radiative activity in the visible region and almost none in the infrared. This may be a fortunate circumstance for life on earth because these two gases account for about 99% of the total atmosphere.

As shown in the figure, triatomic structures have three vibrational modes designated ν_1 for symmetric stretch, ν_2 for bending motion, and ν_3 for antisymmetric stretch. The linear molecule CO_2, lacking a permanent dipole moment, has no oscillating moment and hence no radiative activity in the mode ν_1. However, it does acquire oscillating moments in the modes ν_2 and ν_3, as may be visualized from the figure. The permanent moments of the bent triatomic molecules H_2O and O_3 have oscillating moments in each of the three modes. The CH_4 molecule belongs to the *spherical top* category and has four modes. Although CH_4 has no permanent moment, it acquires oscillating moments in its ν_3 and ν_4 modes. Table 4.3 lists the fundamental vibration frequencies, wavelengths, and wavenumbers of transitions for each mode of the atmospheric species. Because an oscillating moment is

TABLE 4.3 VIBRATIONAL FREQUENCIES, WAVELENGTHS, AND WAVE-NUMBERS OF ATMOSPHERIC RADIATIVE MOLECULES

Species	Parameter	Vibrational Modes ν_1	ν_2	ν_3
CO	Hz	6.43×10^{13}	—	—
	μm	4.67	—	—
	cm^{-1}	2143	—	—
CO_2	Hz	—	2.00×10^{13}	7.05×10^{13}
	μm	—	15.0	4.26
	cm^{-1}	—	667	2349
N_2O	Hz	3.86×10^{13}	1.77×10^{13}	6.67×10^{13}
	μm	7.78	17.0	4.49
	cm^{-1}	1285	589	2224
H_2O	Hz	1.10×10^{14}	4.79×10^{13}	1.13×10^{14}
	μm	2.73	6.27	2.65
	cm^{-1}	3657	1595	3776
O_3	Hz	3.33×10^{13}	2.12×10^{13}	3.13×10^{13}
	μm	9.01	14.2	9.59
	cm^{-1}	1110	705	1043
NO	Hz	5.71×10^{13}	—	—
	μm	5.25	—	—
	cm^{-1}	1904	—	—
NO_2	Hz	3.92×10^{13}	2.26×10^{13}	4.86×10^{13}
	μm	7.66	13.25	6.17
	cm^{-1}	1306	755	1621
CH_4	Hz	8.75×10^{13}	4.60×10^{13}	9.06×10^{13}
	μm	3.43	6.52	3.31
	cm^{-1}	2917	1534	3019
CH_4	Hz	ν_4 5.71×10^{13}		
	μm	5.25		
	cm^{-1}	1904		

Source: Herzberg and Herzberg (1957) and Shimanouchi (1967a, b; 1968).

effective for rotational as well as vibrational transitions, each of the frequencies in the table is the center of a vibrational–rotational band.

All of the vibrational modes available to a polyatomic molecule can be active concurrently. In general, the total vibrational energy is given by an extension of Eq. (4.34) to cover all modes,

$$\mathscr{E}_{tot} = \left(v_1 + \frac{1}{2}\right) h \nu_1 + \left(v_2 + \frac{1}{2}\right) h \nu_2 + \left(v_3 + \frac{1}{2}\right) h \nu_3 + \ldots \quad (4.43)$$

where v_1, v_2, v_3 are the quantum numbers of the modes and ν_1, ν_2, ν_3 are the vibrational frequencies. The energy value of each mode is readily found from Table 4.3 for any value of v, to the extent that anharmonicity does not reduce the nominally equal spacing of energy levels. The evaluation is made by first noting that the listed energy value corresponds to $\Delta\mathscr{E}$ of Eq. (4.39) whereas the energy $\mathscr{E}_v$ of a vibration is given by Eq. (4.34). The two expressions differ only by the factor $v + \frac{1}{2}$. Then multiplication of an energy value in the table by this factor, when v has been selected, gives the energy of the corresponding level of that mode.

The vibrational state is designated by a *term symbol,* which is an ordered listing of quantum numbers for the modes. For example, the symbol 010 means that the quantum number is zero for the v_1, unity for the v_2, and zero for the v_3 mode. When the mode quantum numbers are small, the selection rules are $\Delta v = 0, \pm 1$ with a proviso that one of the numbers shall change. If the vibration is simple harmonic, only one of the mode numbers can change for a given transition. The selection rules are modified, however, by the anharmonicity discussed previously. Simultaneous changes of two mode quantum numbers now can occur, creating sum and difference frequencies. Such a combination frequency usually creates another vibration–rotation band that complicates the spectral picture. Transitions are indicated by writing the term symbol for the upper level first, and then joining the symbols with an arrow. Thus $001 \rightarrow 100$ means an emissive transition in which the quantum number changes from zero to one for the v_1 mode, remains at zero for the v_2 mode, and goes from one to zero for the v_3 mode. This particular transition is used in the CO_2 laser to produce an output at 10.6 μm, which can be roughly verified from data in Table 4.3 if it is noted that no allowance is being made for the small energy contribution of the rotational transition.

A more-detailed view of vibrational energy must include additional factors not covered here but treated in spectroscopy textbooks. However, we mention one of the factors because it appears frequently in the literature of gas lasers that employ linear triatomic species. The v_2 or bending mode of such a structure, indicated in Figure 4.14, has a twofold degeneracy comprising modes v_{2a} and v_{2b}. These motions may be considered as parallel and normal, respectively, to the plane of the diagram. Depending on the phases of these orthogonal motions, the nuclear masses can move in elliptical orbits, and the resulting angular momentum adds to, or subtracts from, that due to rotation of the molecule as a unit. This angular momentum of vibration, defined by quantum number l, effectively couples the vibrational and rotational motions whenever $l \neq 0$. This causes a splitting of rotational lines (degeneracy) and is called *l-type doubling.* The value of l is written as a superscript to the quantum number for the v_2 mode in the term symbol. For example, $00^01 - 10'0$ means that l is zero in the initial level of the transition and unity in the final level.

4.4. ELECTRONIC ENERGY AND TRANSITIONS

Electronic energy is closely related to vibrational energy since both forms derive from the elastic valence bonds that hold the constituent atoms into a molecular entity. Here we review the origins of electronic energy and the resulting electronic states of molecules. Attention is given to the elaborate symbolism necessary to describe the electronic states. Next are described the ways in which the angular momenta of electronic orbital motion and electronic spin can combine with the angular momentum of the nuclear rotation considered in Section 4.2.1. The section concludes with a review of the quantum selection rules commonly encountered.

4.4.1. Origins of Electronic Energy

Electronic energy, which is potential in form, originates in the unstable configurations assumed by the electrons of atoms and molecules. In wave mechanic theory these electrons are characterized by wave functions which, with the corresponding electron motions, are called *orbitals*. Electrons of the inner orbitals, tightly bound to the atomic nucleus, can be disturbed or dislodged only by photons having the large energies of short-wave ultraviolet and X rays. Electrons of the outermost orbitals, less tightly bound and more easily disturbed, are the ones principally involved in absorption–emission in the visible and ultraviolet.

Let two atoms be brought sufficiently close that their outer orbitals interpenetrate. If the atoms rearrange these orbitals to involve both nuclei, they combine into a molecule. These electrons now are shared, and we may regard them as the valence electrons that bind the nuclei together. A stable molecule is formed if the total potential energy of the combination has a minimum value at a particular internuclear distance. This is the equilibrium distance, indicated for each of the potential energy curves in Figure 4.15. At separations less than the equilibrium distance the repulsion forces between the nuclei become predominant. At greater separations, *within limits,* the attraction forces between the shared electrons and the nuclei become predominant and provide the elastic binding force. In Figure 4.15 the curves of electronic potential energy express the same relationship as the curve in Figure 4.12. These two representations, although designated by different names for organization of the subject, refer to the same electronic binding force.

The lower curve of Figure 4.15 represents the energy of an electronic ground state, which was the condition for previous discussions of rotational and vibrational energy. Absorption of a high-energy photon changes the electron configuration to one that has potential energy even when the nuclei are at equilibrium distance. This is a transition to an excited state whose potential energy is represented by the upper curve. The energy gain due to the transition is given by the ordinate difference between the two minima and typically is a few electron volts. This is considerably greater, however, than the energies of the vibrational levels in Figure 4.13. Usually the electron configuration for the upper state exerts a lesser binding force; the force constant becomes less, and the internuclear distance increases accordingly. This slightly reduces the natural vibration frequency, as may be seen from Eq. (4.32). Also, the increased internuclear distance means a greater moment of inertia, hence a decreased rotational constant and a decreased spacing of rotational lines according to Eq. (4.18).

4.4.2. Electronic States of Molecules

Molecular electronic states can be described most easily for a diatomic molecule such as O_2. The symbolism for describing these states is about the

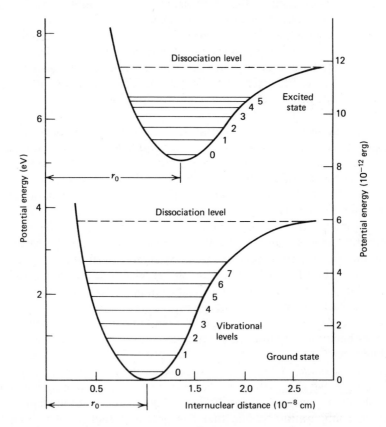

Figure 4.15. Potential energy curves for two electronic states of a diatomic molecule.

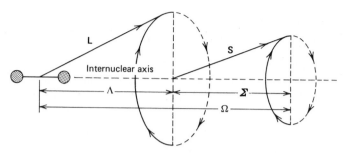

Figure 4.16. Projections of orbital and spin angular momenta on the internuclear axis.

same as for atomic states except that capital rather than lowercase letters are used. Referring now to Figure 4.16 the resultant angular momentum due to the orbital motions of all the electrons is represented by the vector **L**. Its quantized magnitude is defined by

$$\left| \mathbf{L} \right| = \frac{h}{2\pi} [L(L+1)]^{1/2}, \qquad L = 0, 1, 2, \ldots \qquad (4.44)$$

where L is the quantum number for orbital angular momentum. In a molecule such as O_2 having identical nuclei the electrostatic field of the nuclei is symmetrical about the internuclear axis. When the orbital motions are strongly coupled to this field, **L** precesses about the axis in the manner of a gyroscope, as shown.

The axial component of **L** is designated Λ and is quantized into integer multiples of $h/2\pi$. This is analogous to the spatial quantization of angular momentum illustrated in Figure 4.7. Here the value of the axial component is $M_L(h/2\pi)$, with M_L taking any of the values

$$M_L = L, L-1, L-2, \ldots, -L \qquad (4.45)$$

If the orbital motions of all the electrons are reversed, the energy of the molecule does not change but the vector sense of **L** is reversed. This means that M_L becomes $-M_L$. Therefore, these states differing only in the sign of M_L have the same energy and are degenerate. For each value of L there are $L+1$ possible states having *different* energy values. Hence the quantum number Λ of the axial component Λ of orbital angular momentum **L** is defined by

$$\Lambda = \left| M_L \right| = 0, 1, 2, \ldots, L \qquad (4.46)$$

so that Λ can assume only zero or positive values. The vector Λ then has the magnitude

$$\left| \mathbf{\Lambda} \right| = \Lambda \frac{h}{2\pi} \qquad (4.47)$$

A description of electronic states must also take into account the resultant angular momentum of all the electron spins. Representing this resultant by the vector $\mathbf{S}$ as in Figure 4.16, its quantized magnitude is defined,

$$\left| \mathbf{S} \right| = \frac{h}{2\pi} [S(S + 1)]^{1/2} \qquad (4.48)$$

where S is the quantum number for spin. Since the spin of a single electron has associated with it the factor $\frac{1}{2}$, the value of S is half-integral when the total number of electrons is odd and integral when the total number is even. Therefore, S takes one of the values $\frac{1}{2}$, 1, $1\frac{1}{2}$, ... For all states of orbital angular momentum except $L = 0$, the orbital motions produce a magnetic field directed along the internuclear axis. In some cases this field will couple with the electron spin, causing spin vector $\mathbf{S}$ to precess about the internuclear axis, as in the figure. This precession is analogous to that of the orbital angular momentum vector $\mathbf{L}$ about this axis.

The axial component of $\mathbf{S}$ is designated $\mathbf{\Sigma}$, quantized into multiples of $h/2\pi$ analogously to Λ described above. The component vector $\mathbf{\Sigma}$ has the magnitude

$$\left| \mathbf{\Sigma} \right| = \Sigma \frac{h}{2\pi} \qquad (4.49)$$

where Σ is the quantum number. This sloping Σ must not be confused with the upright Σ used for an electronic term symbol, as described below and shown in Table 4.4. Allowed values of Σ are

$$\Sigma = S, S - 1, S - 2, \ldots, -S \qquad (4.50)$$

so that $2S + 1$ values are possible. Evidently Σ can be positive or negative whereas Λ is positive only, as shown by Eq. (4.46). The quantity $2S + 1$ is the *multiplicity factor for spin* and is analogous to the degeneracy factor $2J + 1$ for rotational levels. The multiplicity means that any given state of Λ except $\Lambda = 0$ is split into multiplets known as singlet, triplet, quintet, ... , for $S = 0$, 1, 2, ... , as listed in Table 4.4.

The electronic state of the molecule is symbolized by the *electronic term symbol*, which embodies the foregoing information plus additional factors. The term symbol is Σ, Π, Δ, Φ to represent, respectively, the condition $\Lambda = 0$, 1, 2, 3 as listed in Table 4.4. The value of multiplicity $2S + 1$ is denoted by a preceding superscript to the term symbol. For example, $^3\Sigma$ means that the quantum number $\Lambda = 0$ and that $2S + 1 = 3$. This may seem

TABLE 4.4 NOTATION FOR ELECTRONIC QUANTITIES AND STATES

Quantity	Vector	Designation of States When Quantum number Λ has the Value		
		0	1	2
Orbital angular momentum	$\mathbf{L}$			
Projection on axis	Λ			
Spin angular momentum	$\mathbf{S}$	Σ	Π	Δ
Projection on axis	Σ			
		Designation of Multiplicity $(2S+1)$ When Quantum Number S has the Value		
Sum of Λ and Σ (See Figure 4.16)	Ω			
Total angular momentum (See Figure 4.17)	$\mathbf{J}$	0 ½ 1 1½ 2		
		Single Doublet Triplet Quartet Quintet		

misleading because when $\Lambda = 0$ there is no magnetic field to couple the electron spin to the internuclear axis. In such a case splitting or degeneracy of the Λ state cannot occur. Nevertheless, for all values of Λ the quantity $2S + 1$ is considered the multiplicity for spin.

The total electronic angular momentum about the internuclear axis is the sum of the orbital and spin components and is denoted by Ω, as in Figure 4.16,

$$\Omega = \Lambda + \Sigma \qquad (4.51)$$

The quantum number of this resultant is

$$\Omega = \Lambda + \Sigma \qquad (4.52)$$

Since Λ is integral but Σ is half-integral or integral, depending on whether the total number of electrons is odd or even, Ω is also half-integral or integral. When the value of Ω is to be indicated by the term symbol, it is written as a following subscript.

Additional quantities involving symmetry properties are required to specify an electronic state. In a diatomic or linear triatomic molecule, any plane containing the internuclear axis is an axis of symmetry. Considering a Σ state which is nondegenerate because $\Lambda = 0$, the eigenfunction either remains unchanged in sign, or changes sign, when reflected at such a plane. This property is indicated by a plus or minus sign, written as a following superscript to the term symbol. Somewhat similarly, a homonuclear diatomic molecule such as O_2 has a center of symmetry lying at the midpoint of the internuclear axis. The eigenfunction either remains unchanged in sign, or changes sign, when reflected at this center. In the first case, the state is even, and in the second, the state is odd. These are indicated, respectively, by the letter g or u (from the German *gerade* and *ungerade*) written as a following subscript to the term symbol. For example the ground state of O_2 has the term symbol $^3\Sigma^-_g$, which means that $\Lambda = 0$ and $S = 1$, and that the symmetry properties are $(-)$ and (g). The ground states of N_2 and CO are denoted, respectively, by $^1\Sigma^+_g$ and $^1\Sigma^+$, which mean that for each molecule the spin S and orbital angular momentum L are zero. This agrees with the treatment of rotational energy in Section 4.2.1, where we assumed an electronic ground state so that the angular momentum was due entirely to rotation of the nuclear masses about their center of gravity.

4.4.3. Coupling of Angular Momenta

A molecule has four sources of angular momenta:

Orbital motions of electrons
Spin of electrons
Spin of nuclei
Rotation of molecule as a unit

Angular momentum of nuclear spin is often disregarded, as is done here. The total angular momentum is then the resultant of the remaining three, each of which has been discussed to this point. These momenta are coupled in varying degrees with each other and with the internuclear axis. Five cases of such coupling have been investigated by Hund, and are treated in detail by Herzberg (1961, 1966, 1971), Kondratyev (1965), and Walker and Straw (1967). The coupling cases represent idealized or limiting situations, but many intermediate cases are found in practice. A real molecule may go from one case to another as parameters change. Here we discuss only the frequently occurring cases (a) and (b), which are sufficient to reveal the interactions of electronic and pure rotational momenta.

Hund's case (a), shown in Figure 4.17a, obtains when the electron orbital and spin motions are strongly coupled to the internuclear axis, as described in the preceding section. The component vectors Λ and Σ along this axis are then well defined; each component is quantized into multiples of $h/2\pi$. Because $\Lambda \neq 0$ for this condition, case (a) cannot apply to an electronic ground state whose term symbol is $^1\Sigma$, meaning that $\Lambda = 0$ and $\Sigma = 0$. However, when these components are well defined, the sum vector Ω also is well defined and couples with vector N, which represents the angular momentum of molecular rotation. The resultant vector J, representing *total* angular momentum of the molecule, remains fixed in space. The vectors Ω and N nutate about J, somewhat in the manner that a tilted spinning top moves around a vertical axis. The quantum number J of the resultant vector J can assume one of the values

$$J = \Omega, \; \Omega + 1, \; \Omega + 2, \; \ldots \tag{4.53}$$

Since Ω is either half-integral or integral, depending on whether the total number of electrons is odd or even, the same stipulation applies to J.

Hund's case (b) is illustrated in Figure 4.17b. This case obtains when, although we may have $\Lambda \neq 0$, the magnetic field from the orbital motions is not sufficient to couple the spin vector $\mathbf{S}$ to the internuclear axis. Naturally it must also obtain when $\Lambda = 0$, meaning that there is no magnetic field at all. Such a condition is represented by the term symbol Σ, as shown in Table 4.4. In either event there is no axial component Σ of $\mathbf{S}$ so that the sum vector Ω is not defined. When $\Lambda \neq 0$, the corresponding vector Λ couples with $\mathbf{N}$ to form the resultant $\mathbf{K}$, whose quantum number K can assume any of the integral values

$$K = \Lambda, \Lambda + 1, \Lambda + 2, \ldots \tag{4.54}$$

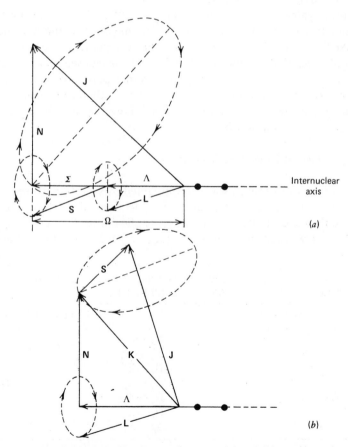

Figure 4.17. Hund's coupling cases (a) and (b).

The spin vector $\mathbf{S}$ combines with $\mathbf{K}$ to form the resultant $\mathbf{J}$, which remains fixed in space while the components precess about it. The quantum number of $\mathbf{J}$ can assume any of the values

$$\mathbf{J} = (K + S), \; (K + S - 1), \; (K + S - 2), \; \ldots \tag{4.55}$$

according to the rules for adding two quantized vectors. Because S is either half-integral or integral, the same stipulation applies to J.

Case (b) represents the situation for the discussion of rotational energy in Section 4.2.1 where an electronic ground state was assumed. Since the term symbol for the ground state is Σ, this means that the quantum number Λ and the representing vector Λ for the axial component of orbital angular momentum are zero. For this condition, vectors $\mathbf{N}$ and $\mathbf{K}$ are identical, as may be visualized from Figure 4.17b. We note a further relevance with the earlier discussion that employed the CO molecule for numerical examples. The term symbol for the ground state of CO is $^1\Sigma^+$, indicating that the multiplicity $2S + 1$ for spin is unity. This means that the spin quantum number $S = 0$, so that from Eq. (4.48) the corresponding vector $\mathbf{S} = 0$. For this double restriction, that $\Lambda = 0$ and $S = 0$, we can see that the vectors $\mathbf{K}$ and $\mathbf{N}$ coincide into $\mathbf{J}$, which thereby becomes perpendicular to the internuclear axis. The ground electronic state $^1\Sigma^+$, therefore, can contribute nothing to the angular momentum, which is now due entirely to the rotation of the molecule as a unit body.

4.4.4. Selection Rules for Transitions

Electronic transitions, always accompanied by vibrational and rotational transitions, are governed by numerous selection rules. Here attention is called to the general rules having widest applicability. The resulting allowed transitions are listed in Table 4.5 without regard to vibrational and rota-

TABLE 4.5 ALLOWED ELECTRONIC TRANSITIONS FOR DIATOMIC MOLECULES HAVING THE SAME MULTIPLICITIES

Unequal Nuclear Charge	Equal Nuclear Charge	
$\Sigma^+ \leftrightarrow \Sigma^+$	$\Sigma_g^+ \leftrightarrow \Sigma_u^+$	
$\Sigma^- \leftrightarrow \Sigma^-$	$\Sigma_g^- \leftrightarrow \Sigma_u^-$	
$\Pi \leftrightarrow \Sigma^+$	$\Pi_g \leftrightarrow \Sigma_u^+$	$\Pi_u \leftrightarrow \Sigma_g^+$
$\Pi \leftrightarrow \Sigma^-$	$\Pi_g \leftrightarrow \Sigma_u^-$	$\Pi_u \leftrightarrow \Sigma_g^-$
$\Pi \leftrightarrow \Pi$	$\Pi_g \leftrightarrow \Pi_u$	
$\Pi \leftrightarrow \Delta$	$\Pi_g \leftrightarrow \Delta_u$	$\Pi_u \leftrightarrow \Delta_g$
$\Delta \leftrightarrow \Delta$	$\Delta_g \leftrightarrow \Delta_u$	

Source: From Herzberg (1961).

tional factors. In indicating a transition the term symbol for the higher state usually is written first, thus $\Pi \leftrightarrow \Sigma$. The double arrow means that the transition can proceed in either direction, while $\nleftrightarrow$ means that the transition is forbidden. The direction of a single arrow, say toward or from the higher-state term symbol, shows whether absorption or emission is being considered. In all transitions an even (gerade) and an odd (ungerade) state are involved, that is, transitions obey the rules $g \leftrightarrow u$, $g \nleftrightarrow g$, $u \nleftrightarrow u$.

Several quantum selection rules are noted. The quantum number Λ of the axial component of orbital momentum can change according to

$$\Delta \Lambda = 0, \ \pm 1 \tag{4.56}$$

As shown in Table 4.5, this rule means that transitions such as

$$\Sigma \leftrightarrow \Sigma, \ \Pi \leftrightarrow \Sigma, \ \Pi \leftrightarrow \Pi, \ \Delta \leftrightarrow \Pi$$

can occur but $\Delta \leftrightarrow \Sigma$ is forbidden. For the quantum number S of electron spin, as used in Eqs. (4.48) and (4.50), the rule is

$$\Delta S = 0 \tag{4.57}$$

Hence from Eq. (4.50)

$$\Delta \Sigma = 0 \tag{4.58}$$

Since the multiplicity of a state is defined by the factor $2S + 1$, the rule (4.57) means that transitions between states of different multiplicities are forbidden.

Changes in the quantum number J of total angular momentum are governed by

$$\Delta J = 0, \ \pm 1 \tag{4.59}$$

with the restriction that $\Delta J = 0$ is forbidden when $\Omega = 0$ in *both* electronic states. If both states of the transition belong to coupling case (*b*), changes in the quantum number N of the vector **N** obey the rule

$$\Delta N = 0, \ \pm 1 \tag{4.60}$$

with the restriction that $\Delta N = 0$ is forbidden for Σ—Σ transitions. It is evident from Figure 4.17 that the foregoing selection rules for ΔJ and ΔN may involve changes in the quantum numbers of molecular angular momentum because vector **N** is a component of **K** and **J**. Such changes correspond to rotational transitions and determine the fine structure of electronic band systems.

The entire body of selection rules, only a few of which are cited above, simplify the treatment of electronic spectra. However, not all of the rules have absolute validity throughout the total range of phenomena. Many exceptions called *forbidden transitions* occur and they usually produce weaker spectra than normal. Herzberg (1961) states the matter succinctly:*

Forbidden transitions can be observed in absorption by using very long absorbing paths (considerably longer than are necessary for the ordinarily allowed transitions). In emission they appear only under quite special conditions of excitation. Apart from the weak occurrence of whole band systems in violation of some electronic selection rule, certain branches that are forbidden by the ordinary selection rules may appear very weakly in the bands of an otherwise non-forbidden band system. The occurrence of forbidden transitions may have one of the three following reasons:

1. The selection rule that is violated may hold only to a first approximation.

2. The selection rule may hold strictly for dipole radiation but not for quadrupole radiation or magnetic dipole radiation.

3. The selection rule may hold only for the completely free and uninfluenced molecule, and may be violated in the presence of external fields, collisions with other molecules, and the like.

The electronic states available to molecules as a result of the selection rules are presented conveniently on energy-level diagrams similar to those employed for atoms. A simplified diagram listing a few states of the CO molecule is given in Figure 4.18. Vibrational levels frequently are shown as in the figure to enhance the energy picture; usually it is not feasible to draw the closely spaced rotational levels. Electronic transitions are indicated by the arrows between states. Each electronic transition, and the accompanying vibrational and rotational transitions, produce an electronic band system. The system is usually named for the discoverer or early investigator, as indicated in the figure. Band systems are considered in the following chapter, but at this point we note that they are quite numerous. For example, the CO molecule is one of the simplest species in the atmosphere, but 29 types of transitions are found among its 23 electronic states. Krupenie (1966) describes 23 of the resulting band systems, all of which lie between 60 and 860 nm ($= 0.86$ μm).

The degree of electronic excitation has a marked effect on the vibration potential curve. This can be appreciated from Figure 4.19, which shows the ground and three excited states of the O_2 molecule. Each state accommodates a number of vibration levels, but the accommodation decreases in higher states because the dissociation limit is reached sooner. In effect, the higher states are shallower wells of potential. The decreased force constant of the valence bond in an excited state means that the internuclear equilibrium distance is greater than in the ground state. This is seen, for example, by comparing the potential curves of the $^3\Sigma_g^-$ and $^3\Sigma_u^-$ states in the figure. The decreased force constant means a decreased vibrational frequency, from Eq. (4.42). The greater equilibrium distance produces a greater moment of inertia and from Eq. (4.13) a smaller value of the rotational constant. All such factors affect the structures of band systems.

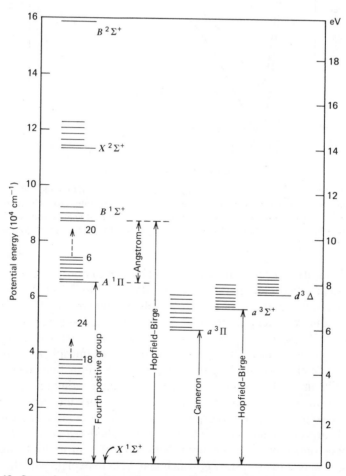

Figure 4.18. Some electronic states and corresponding energy values of a CO molecule. After Herzberg (1961). *Molecular Spectra and Molecular Structure, Vol. I, Spectra of Diatomic Molecules.* Copyright © 1950 by Van Nostrand Reinhold Company. Reprinted by permission of Van Nostrand Reinhold Co. And Krupenie, (1966).

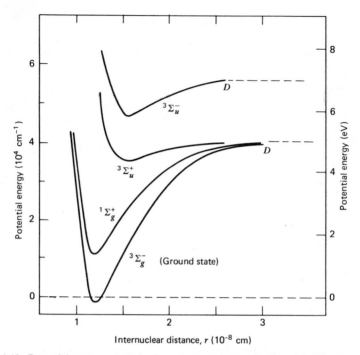

Figure 4.19. Potential energy curves for four electronic states of an O_2 molecule. The vertex of each curve is at the internuclear equilibrium distance for that state. From Herzberg (1961). *Molecular Spectra and Molecular Structure, Vol. I. Spectra of Diatomic Molecules.* Copyright © 1950 by Van Nostrand Reinhold Company. Reprinted by permission of Van Nostrand Reinhold Co.

The $^3\Sigma_u^-$ state indicated in Figure 4.19 has great importance for absorption of solar ultraviolet energy. The transitions $^3\Sigma_u^- \leftrightarrow {}^3\Sigma_g^-$ involve energies, and hence wavenumbers, of approximately 50,000 cm^{-1} and greater. The corresponding wavelengths are near 200 nm and less, which is well into the ultraviolet. These transitions produce the Schumann–Runge bands, sufficiently strong to make even the rarefied upper atmosphere completely absorbing in this spectral region. At the dissociation limit near 56,850 cm^{-1} = 176 nm the bands merge into an absorption continuum that is effective in producing atomic oxygen, some of which goes into the creating of ozone. The relatively low-energy transition $^1\Sigma_g^+ \leftrightarrow {}^3\Sigma_g^-$ is responsible for weak absorption bands near 700 nm often called *red bands*. This is a forbidden transition on two counts, between different multiplicities, and from even state g to even state g, which may help to explain why the effects are weak.

5

Spectra of Energy Transitions

Spectra of absorption and emission are the observables of energy transitions and thus constitute the basic data upon which spectroscopy is built. In this chapter we review the organization rules for spectra and show the methods of constructing schematics for transitions and spectra. Pure rotational spectra are considered first followed by the vibration–rotation spectra that have great importance in the infrared. Vibrational–electronic transitions are then discussed in detail; the meaning of the Franck–Condon principle is illustrated; and the rotational structure of electronic bands is explained. The chapter concludes with a look at the types of transitions that produce fluorescence and phosphorescence.

5.1. ROTATIONAL SPECTRA

Pure rotational spectra occur only in the extreme infrared and microwave regions because of the small energy value of a rotational transition. This value is defined by Eq. (4.17a), repeated here for reference,

$$\Delta \mathscr{E} = 2BhcJ' \tag{5.1}$$

based on the quantum selection rule $\Delta J = \pm 1$. The spectral parameters of the rotational lines are defined by Eqs. (4.18), repeated here,

$$\nu = 2B c J' \quad \text{(Hz)}$$

$$\lambda = \frac{1}{2BJ'} \quad \text{(cm)}$$

$$\bar{\nu} = 2BJ' \quad (\text{cm}^{-1}) \tag{5.2}$$

where J' can take any value 1, 2, 3. . . .

Transition energy values and spectral parameters for the CO molecule are combined in the display of Figure 5.1. The lowest transition that can

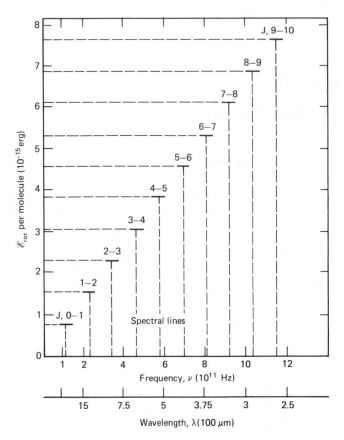

Figure 5.1. Rotational transition energy $\mathcal{E}_{rot}$ and spectral parameters of absorption–emission lines for CO molecule.

occur is between $J'' = 0$ and $J' = 1$, and produces a line at $\lambda = 0.259$ cm = 2.59 mm. Correspondingly, we have the very low value of wavenumber $\tilde{\nu} = 3.86$ cm^{-1} for this transition. Interestingly, this is the wavelength emitted by the tenuous but vast clouds of CO gas within the Milky Way, and astronomers use radio telescopes at this wavelength to probe the galactic structure. In such interstellar space a CO molecule makes the transition $J'' = 0$ to $J' = 1$ either by absorption or by collision with another molecule. Although the individual collision probability may seem very small, there are large amounts of space and time to balance the scale. In emitting this 2.59-mm wavelength, the molecule simply stops rotating, thereby dropping back to $J'' = 0$. An interesting account of galactic carbon monoxide is given by Gordon and Burton (1979).

5.2. VIBRATION–ROTATION BANDS

Molecular vibration produces an oscillating electric dipole moment that is effective for both vibrational and rotational transitions. Hence both types occur together, and the resulting energy quantum is equal to the sum of the separate transition energies. Because the energy of the vibrational transition is much greater than that of the rotational, as may be seen by comparing the levels in Figures 4.5 and 4.13, and because many rotational levels are active, the spectrum of the combined transitions is an array of rotational lines grouped around the vibrational frequency. The energy values are such that this *vibration–rotation band* lies in the infrared at wavelengths less than about 20 μm. A pure vibration spectrum of a gas seldom can be observed. The effects of rotation can be avoided only by placing the substance in solution, so that rotations are suppressed by the dense molecular packing that is characteristic of a liquid.

Vibration–rotation bands are the marks of gases whose molecules are in the electronic ground state. Internal energy is then equal to the sum of its rotational (4.14) and vibrational (4.36) energies,

$$\mathscr{E}_{J,v} = BhcJ(J+1) + \left(v + \frac{1}{2}\right)\frac{h}{2\pi}\left(\frac{K}{m'}\right)^{1/2} \tag{5.3}$$

where *erg* is the energy unit. This can be put more concisely by employing the term values (4.15) and (4.35),

$$\mathscr{E}_{J,v} = F(J) + G(J) = BJ(J+1) + \left(v + \frac{1}{2}\right)\omega' \tag{5.4}$$

where cm^{-1} is the energy unit and ω' is the vibrational frequency in cm^{-1}. In simple cases the quantum selection rule is $\Delta v = \pm 1$, except that $\Delta v = -1$ cannot apply to $v = 0$. The ratio of population in $v = 1$ to that in $v = 0$ is only 3.26×10^{-5} at 300 K, as found in Eq. (4.42). In the atmosphere, therefore, the band will be observed more readily in absorption than in emission, although this circumstance is not essential to this discussion. The rotational transitions are considered next. For diatomic molecules, and linear triatomic molecules in the stretching modes v_1 and v_3, the change of dipole moment occurs parallel to the internuclear axis, as may be understood from Figure 4.14. For this condition the selection rule is $\Delta J = \pm 1$.

Considering absorption by a diatomic species in the *electronic ground state,* the selection rules are

$$\Delta v = 1 \qquad \Delta J = \pm 1$$

Most of the molecules are in the level $v'' = 0$ but are distributed over numerous levels of J. Hence any molecule in going to the level $v' = 1$ can go either to the next-higher rotational level, for $\Delta J = 1$, or to the next-lower level, for $\Delta J = -1$, with about equal probabilities. In the first instance the overall transition energy is increased by the rotational increment, while in the second it is decreased. This is expressed by the algebraic sum of Eqs. (4.17a) and (4.39),

$$\Delta \mathscr{E}_{J,v} = \pm 2Bhc J' + \left(\frac{h}{2\pi}\right)\left(\frac{K}{m'}\right)^{1/2} \tag{5.5}$$

where J' is the quantum number in the upper vibrational level. As with the transition energy, so with the spectral frequency of the line, which is found by substituting Eq. (5.5) into Eq. (1.14) to give

$$\nu = \pm 2BcJ' + \left(\frac{1}{2\pi}\right)\left(\frac{K}{m'}\right)^{1/2} \tag{5.6}$$

This result is equal, of course, to the sum of Eqs. (4.18a) and (4.40).

Because many closely spaced rotational levels are involved and because many molecules are present in a volume of gas, the numerous transitions create a band of rotational lines grouped on each side of the vibrational frequency. From the relation $\nu = c\tilde{\nu}$ the wavenumber of each line is

$$\tilde{\nu} = \pm 2BJ' + \left(\frac{1}{2\pi c}\right)\left(\frac{K}{m'}\right)^{1/2} \tag{5.7}$$

which can also be obtained from the term values (5.4) for two successive values of J. It is clear from Eq. (5.7) that the wavenumber spacing of adjacent lines is just $2B$ cm^{-1}, as for a pure rotation spectrum. Again we see the relative simplicity of designating spectral location by the cm^{-1} unit.

Several of the simultaneous transitions available to diatomic molecules and to linear triatomic molecules in modes ν_1 and ν_3 are shown schematically in Figure 5.2. The vibrational transition obeys the selection rule $\Delta v = 1$, so that the overall result is *absorption*. The arrows at the left of the dotted vertical line correspond to rotational transitions following the selection rule $\Delta J = -1$. Since the value of J at each arrowhead of this group is less by one than the value at the tail, this rotational energy subtracts from the vibrational. This group is the lower-energy, hence lower-frequency, portion of the band and is called the *P-branch*. The group of arrows at the right corresponds to transitions following the rule $\Delta J = 1$; this is the higher-energy, higher-frequency *R-branch*.

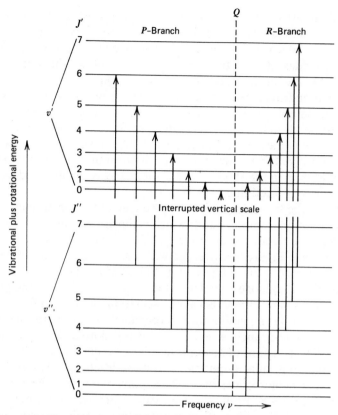

Figure 5.2. Simultaneous transitions in vibrational and rotational energies.

The progressive increase of arrow length in going to the right indicates that the rotational energy transitions are increasing. This occurs for decreasing values of J'' in the P-branch and for increasing values of J' in the R-branch. The rotational level spacings in the level v' are somewhat smaller than those in the level v'', as shown by exaggerated scale in the figure. This is a result of the increased moment of inertia in higher vibrational levels, as pointed out in the discussion of Figure 4.12, and the consequent reduction of rotational energy according to Eq. (4.12). Therefore, the lengths of the arrows do not increase by a constant amount in going to the right, and the frequency spacing of the lines decreases slightly, an effect shown here only qualitatively.

The branches P and R are called *parallel branches* because the dipole moment oscillates parallel to the internuclear axis for the conditions stated above. For such vibrational modes the transition $\Delta J = 0$ is forbidden. Thus no transition arrow and no spectral line appear in Figure 5.2 for this transition, which would be only at the vibrational frequency itself. The situation is otherwise, however, for the vibrational mode v_2 of linear triatomic and the three modes of bent triatomic molecules, all of which are shown in Figure 4.14. For these modes the change of dipole moment has a component perpendicular to an internuclear axis, as may be seen in that figure. The rotational selection rule is now $\Delta J = 0, \pm 1$, which creates a Q-branch that corresponds to $\Delta J = 0$. This branch, known as a *perpendicular branch,* occurs at the vibrational frequency itself. In simple cases it appears as a broad, unresolved line. However, if the moment of inertia is notably less in the level v' than in the level v'', the Q-branch may be observed as a group of very closely spaced lines.

Several of the foregoing factors may be seen in the spectrometer trace of the CO vibration–rotation band, which is illustrated only schematically in Figure 5.2. The trace, centered at approximately $\bar{\nu} = 2144$ cm^{-1} or $\lambda = 4.69$ μm, is shown in Figure 5.3, overlaid with a schematic of transitions for the *absorptive* transitions $v'' = 0$ to $v' = 1$. Values of J as high as 18 or so would be needed to schematize completely the trace that is shown. The Q-branch naturally does not appear because oscillation of the dipole moment is parallel to, but has no component perpendicular to, the internuclear axis. Absence of a Q-branch is indicated by the dotted line separating the branches. The lessening of absorption at lower values of J may be attributed to the lesser molecular populations of these levels, as likewise the lesser absorption at higher values of J. These populations as a function of J are shown in Figure 4.8, from which we expect greatest absorption at levels near $J = 7$ at 300 K. Adjacent absorption peaks, which are separated by $2B$ cm^{-1}, are the centers of individual rotational lines. From the value $B = 1.93$ in Table 4.2, this separation is equivalent to 3.86 cm^{-1}. A slight asymmetry of line spacing, seen here as a bunching of the lines at higher wavenumbers, may just be discerned in reading the figure from left to right. This is due to the slight decrease in rotational-energy-level spacing, shown in exaggerated form by Figure 5.2.

The spectral simplicity of the CO band in Figure 5.3 is a rather special case. While it illustrates very well the main features of a simple band, it does not disclose the complexities usually found in vibration–rotation bands. First, consider the different vibrational modes and frequencies listed in Table 4.3. All of these modes, and corresponding transitions, are active concurrently in a volume of gas containing many molecules. Each modal transition produces a vibration–rotation band having the center frequency listed. Each of these bands has P- and R-branches and a Q-branch on occasion. Frequent overlapping of bands occurs even for the same molecu-

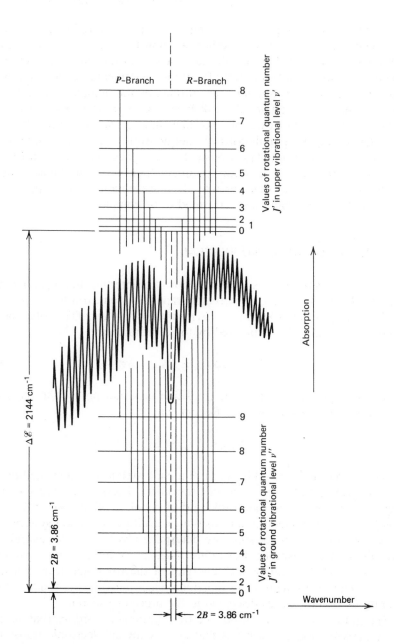

Figure 5.3. Spectrometer trace of the vibration–rotation band of CO, observed in absorption at room temperature. Rotational levels are shown to scale, but the distance between lower and upper vibration levels should be about five times that shown. From Bauman (1962).

lar species, for example, the ν_1 and ν_3 branches of H_2O whose center frequencies are quite close together, as shown in the table. Anharmonic vibrations at higher values of v produce weak overtone bands whose center frequencies are not integer multiples of the fundamental. Combination, or sum and difference, frequencies are formed by the fundamental and overtones, resulting in additional weak bands. Complexities also arise from rotational factors. For example, the lines have greatly different strengths that depend on the populations of energy levels and the probabilities of transitions. Also, as noted previously, the line spacings are not uniform but decrease slightly as the quantum numbers J and v assume higher values. Furthermore, the line spacings may decrease notably in the upper electronic states. Atmospheric spectra tend to be complex.

5.3. ELECTRONIC BAND SYSTEMS

Simultaneous transitions in all three forms of internal energy, apparently initiated by the high-energy, extremely rapid electronic transition, produce spectral arrays of lines known as *electronic band systems*. The total energy absorbed or emitted is expressed by the algebraic sum,

$$\Delta\mathscr{E}_{tot} = \pm\Delta\mathscr{E}_{rot} \pm \Delta\mathscr{E}_{vib} + \Delta\mathscr{E}_{el} \tag{5.8}$$

where the sign $\pm$ means that the transition need not proceed in the same direction as the electronic one. Recognizing that

$$\Delta\mathscr{E}_{rot} < \Delta\mathscr{E}_{vib} \ll \Delta\mathscr{E}_{el}$$

it is clear that the spectral location of a band system is determined by the large energy value of the electronic transition. These values are such that the band system lies in the visible and ultraviolet regions. The totality of bands, for a given electronic transition, constitutes the band system, and band locations within a system are determined by the smaller energies of the vibrational transitions. Locations of the actual lines within a band are determined by the still smaller energies of the rotational transitions. In form but not in frequency, an individual electronic band is similar to the vibration–rotation bands discussed in the preceding section.

The individual lines are produced by rotational transitions that accompany the vibrational transitions that accompany an electronic transition. This is the house that Jack builds. Figure 5.4 shows in simple form the transition possibilities between levels and how the lines are grouped into bands. Because we are not concerned with transitions in the same elec-

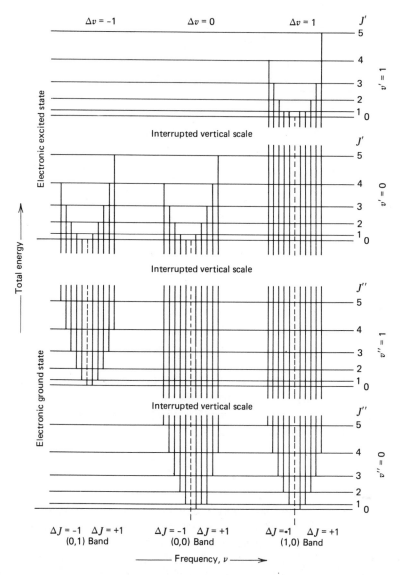

Figure 5.4. Formation of bands by simultaneous transitions in rotational, vibrational, and electronic energies. The schematic applies to a diatomic species for which $\Delta J = 0$ is not allowed.

tronic state, the symbols v'' and v' denote, respectively, a vibrational level in the lower electronic state and a level in the upper state. A band may be designated by writing the final and initial quantum numbers in that order;

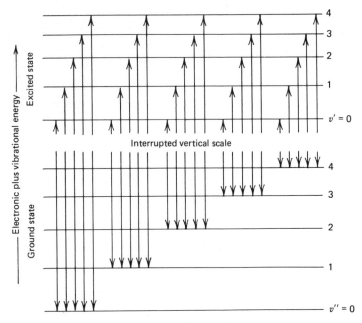

Figure 5.5. Transition progressions between vibrational levels in two electronic states.

for example, the designation (1,0) refers to a band resulting from the transition $v'' = 0$ to $v' = 1$. In the figure the selection rules for ΔJ and Δv are listed beneath each band. From Eq. (5.8) and the relation $v = \mathscr{E}/h$, the frequency of any line is given by

$$v = \frac{\pm \Delta \mathscr{E}_{rot} \pm \Delta \mathscr{E}_{vib} + \Delta \mathscr{E}_{el}}{h} \tag{5.9}$$

Line wavenumber and wavelength can be similarly expressed from the relations in Eq. (1.14). Band systems of the general type shown (simply) in Figure 5.4 exist for all species of gas molecules.

In an infrared vibration–rotation band the molecules *remain* in the electronic ground state, and changes in the vibrational quantum number mostly follow the rule $\Delta v = \pm 1$. Exceptions do occur when the vibrations are strongly anharmonic. With an electronic transition, however, there are no formal rules for changes in v, which can have any value such as

$$\Delta v = 0, \ \pm 1, \ \pm 2, \ \ldots \tag{5.10}$$

All values are not equally probable, as shown by the Franck–Condon principle discussed in Section 5.4. The large number of vibrational transitions that are possible according to Eq. (5.10) can be visualized from Figure 5.5 even though only a few levels are shown. In principle, a transition can take place between any levels v'' and v', and in some cases transitions have been observed at $v = 20$ in electrical discharges where the excitation energy is high. Each vibrational transition in Figure 5.5 produces a band according to the scheme of Figure 5.4, so that a typical system consists of several bands of many rotational lines. In each band the center frequency corresponding to the selection rule $\Delta J = 0$, that is, $J'' = J' = 0$, is called the *band origin*. The detailed structure of a band is described in Section 5.5.

Band systems usually are analyzed in terms of *band progressions,* obtained by organizing the wavenumber data for band origins into a format like that of Table 5.1. Such a format is a *Deslandres table,* and the example here covers a few of the bands produced by the emissive transition $^1\Pi \rightarrow {}^1\Sigma^+$ in CO. These two electronic states are shown in the energy-level diagram of Figure 4.18, and the bands are designated there as the *fourth positive group.* Designations such as positive or negative groups refer to the occurrence of these bands in the positive column or negative glow, respectively, of an electric discharge in a gas at low pressure. The complete tabulation of these band-origin wavenumbers, of which Table 5.1 is a small extract, runs to quantum numbers as high as $v = 20$. Electronic transitions of CO molecules are not particularly important in atmospheric optics, but the wealth of data on this thoroughly studied molecule makes it useful for illustration, as we frequently have seen. At this point it provides a numerical example for showing the vibrational structure of an electronic band system.

TABLE 5.1   WAVENUMBERS OF BAND ORIGINS FOR THE EMISSIVE TRANSITIONS $^1\Pi \rightarrow {}^1\Sigma^+$ OF CO

$v' \diagdown v''$	0	1	2	3	4
0	64,749	62,602	60,485	58,393	56,329
1	66,231	64,088	—	59,882	57,818
2	67,675	65,533	63,416	61,325	—
3	69,088	66,944	64,828	—	60,675
4	70,470	68,323	66,199	64,117	62,055

Source: Herzberg (1961). *Molecular Spectra and Molecular Structure, Vol. I. Spectra of Diatomic Molecules.* Copyright © 1950 by Van Nostrand Reinhold Company. Reprinted by permission of Van Nostrand Reinhold Co.

The principles of *band progressions* may be understood by reviewing together Table 5.1 and Figure 5.5. The transitions whose wavenumbers are listed in the first column of the table all have the lower level $v'' = 0$ in the electronic ground state $^1\Sigma^+$, and upper levels that are progressively $v' = 0$, 1, 2, ... , in the electronic upper state $^1\Pi$. This first column corresponds to the first group of arrows at the left in Figure 5.5 and is a v' progression (v' is increasing) with v'' remaining at 0. Similarly, the second column corresponds to the second group of arrows from the left, and it is a v' progression with v'' remaining at 1, and so on for the other columns and groups of arrows. Hence a v' progression, for a given value of v'', is one in which the transition energy increases, and in which the origins of successive bands occur at progressively higher wavenumbers. This may be seen from the increasing wavenumbers in any one column, and from the increasing arrow lengths in any one group. The decreasing spacing between band origins, in the direction of higher wavenumbers, in any column, is caused by the decreasing spacing of the vibrational levels at higher values of v' in the upper electronic state, as shown in Figure 4.15.

In a similar manner v'' progressions are formed from the rows of Table 5.1. The transitions whose wavenumbers are listed across the top row all have the upper level $v' = 0$ in the upper electronic state $^1\Pi$, and lower levels that are progressively $v'' = 0$, 1, 2, ... , in the lower state $^1\Sigma^+$. This row is a v'' progression (v'' is increasing) with v' remaining at 0, and corresponds to a set of arrows formed by taking the first arrow from each group, going from left to right. The second row of the table is a v'' progression with v' remaining at 1, and corresponds to set of arrows formed by taking the second arrow from each group, going from left to right, and so on for the other rows and arrows. A v'' progression, for a given value of v', is one in which the transition energy decreases, with the successive band origins occurring at progressively lower wavenumbers. The decreasing spacing between band origins, in going across any row, is a result of the decreasing spacing between vibrational levels at higher values of v'' in the lower electronic state, as in Figure 4.15. The effect is not so pronounced as the corresponding one for v' progressions described above.

We should observe that the difference between wavenumbers for any value of v', in the first two columns of Table 5.1, represents the energy difference between the levels $v'' = 0$ and $v'' = 1$, both in the electronic ground state. This is verified by comparing the first arrows from the first two groups of Figure 5.5. This fact allows an interesting comparison, in that a vibrational transition of CO occurs in the infrared, with the molecule remaining in the electronic ground state. Referring now to the table, the wavenumber difference between $v'' = 0$ and $v'' = 1$ at the level $v' = 0$, or even better, the differences between the averages of $v'' = 0$ and $v'' = 1$ at all levels of v', then should equal the wavenumber of the infrared transition between the vibrational levels $v = 0$ and 1. As an example, the difference between the mean values of the first two columns is 2144 cm^{-1}. For comparison, in Table 4.3 the wavenumber of the fundamental vibration of CO is 2143 cm^{-1}. Such close agreement between data gathered in the far ultraviolet and middle infrared regions, using considerably different instrumentation, indicates something of the precision of spectroscopy.

The lack of formal restrictions on changes of v, as shown by Eq. (5.10), means that transitions occur between many levels, such as those listed in Table 5.1. Most all of these have equal probabilities, however, and this fact in conjunction with the generally sparse populations of higher levels may render a particular band so weak that its observation is difficult. The weakness may be in either absorption or emission, or both. Long optical paths, usually possible in the atmosphere, or greater molecular concentrations, usually possible only in the laboratory, often must be used to obtain an optical thickness sufficient for a measurable effect. In the atmosphere, molecular populations of the initial levels are usually the decisive factors of line and band strengths. As found previously, the population of level $v = 1$ is only about 3×10^{-5} that of the level $v = 0$ for a temperature of 300 K. Hence even at the greatest temperatures occurring in the lower atmosphere, most of the transitions must start from the level $v'' = 0$. This means that a v' progression with $v'' = 0$, which corresponds to the group of arrows at the extreme left of Figure 5.5, is far more observable in absorption than in emission. The band strengths then depend on the v' levels at which the transitions end, and these levels are largely determined by the Franck–Condon principle, which we examine next.

5.4. RELATIONS IN VIBRATIONAL–ELECTRONIC TRANSITIONS

Transitions between vibrational levels in different states generally conform to the *Franck–Condon principle,* which can be stated in part as: "The electronic portion of a transition occurs so rapidly (about 10^{-16} sec) compared to the period of a typical vibration (about 10^{-14} sec), that the positions and velocities of the vibrating nuclei do not change appreciably during the transition." Putting it somewhat more in line with probability considerations, the principle asserts that the most probable vibration transition is the one that requires the least alteration of internuclear distance.

Originally the principle assumed that the transition is most probable when the nuclei are at the ends of the vibrational excursion. These are the velocity reversal points where, classically, the nuclei should be found most of the time. This assumption has been modified by conclusions from wave

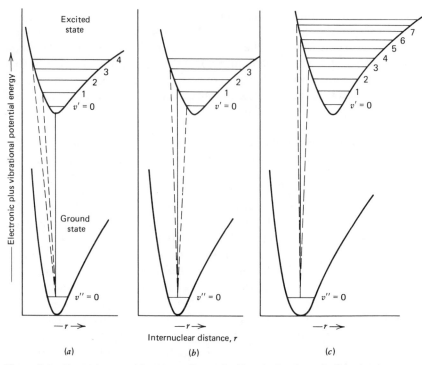

Figure 5.6. Absorptive transitions according to the Franck–Condon principle, for three cases of internunclear equilibrium distance. The most probable transitions follow a vertical line. After Herzberg (1961). *Molecular Spectra and Molecular Structure, Vol. I. Spectra of Diatomic Molecules.* Copyright © 1950 by Van Nostrand Reinhold Company. Reprinted by permission of Van Nostrand Reinhold Co.

mechanics theory, as shown by Figures 4.10 and 4.11 and the discussion following Eq. (4.37). The import there is the idea of probability diffuseness, or inability to predict exactly the positions and velocities (momenta) of the nuclei. This is in line, of course, with the uncertainty principle. Regarding this matter, for purposes here we assume that: (1) At the level $v'' = 0$ the transition has greatest probability of occurrence when the internuclear distance is near the equilibrium value rather than at an extreme value, and (2) at higher levels the transition has good probability when the internuclear distance is *near* an extreme value.

The meaning of the Franck–Condon principle is seen best in its applications. For absorptive transitions the vibrational level of greatest interest is the densely populated level $v'' = 0$. Most of these transitions will start here. Figure 5.6 shows the potential energy curves of the ground and excited electronic states for three equilibrium values of internuclear distance. In

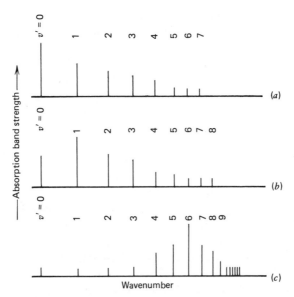

Figure 5.7. Strength distributions in absorption bands for v' progressions, with $v'' = 0$. The three distributions correspond to the transition cases in Figure 5.6. From Herzberg (1961). *Molecular Spectra and Molecular Structure, Vol. I. Spectra of Diatomic Molecules.* Copyright © 1950 by Van Nostrand Reinhold Company. Reprinted by permission of Van Nostrand Reinhold Co.

part (a) the equilibrium distances are equal in the two states, that is, the minimum of the upper curve lies directly over the minimum of the lower. This is similar to the case of the $^1\Sigma_g^+$ and $^3\Sigma_g^-$ states of the O_2 molecule shown in Figure 4.19. In accordance with the eigenfunction of Figure 4.10 and with the Franck–Condon principle, a solid line is drawn vertically upward from the center of the level $v'' = 0$, in the ground state, to the level $v' = 0$ in the excited state. To satisfy the principle strictly, this transition should be the only one that occurs. Because of the probability diffuseness, however, transitions to the levels $v' = 1, 2, \ldots$, actually do occur as indicated by the dashed lines, but with decreasing probability. We thus have a v' progression with $v'' = 0$, and the relative strengths of the resulting absorption bands are illustrated by Figure 5.7a. From the level $v'' = 1$ similar but far fewer transitions take place, because of the relatively sparse population at usual temperatures, creating a second progression of bands, and so on.

In part (b) of Figure 5.6 the equilibrium distance is somewhat greater in the excited state than in the ground state. Again, the most probable transition must be the one that requires the least alteration of internuclear distance. Such a transition is indicated by the solid vertical line from $v'' = 0$ to $v' = 1$. The nuclei originally were near to, or with good probability were at, the equilibrium distance. Because the equilibrium distance is greater in the excited state, immediately after the transition the nuclei are not at, and may not even be near, the equilibrium distance in this level $v' = 1$ of the excited state. Hence they experience a restoring force, and vibration ensues at this level. As the figure is drawn, this transition from $v'' = 0$ to $v' = 1$ is the most probable. Less probable transitions to levels above and below $v' = 3$ do occur, as suggested by the dashed lines. In such a manner a v' progression of bands if formed, and the resulting band strengths are shown in part (b) of Figure 5.7.

In part (c) of Figure 5.6 the equilibrium distance in the excited state is still greater. This is similar to the $^3\Sigma_u^-$ and $^3\Sigma_g^-$ states of the O_2 molecule in Figure 4.19. In obeying the Franck–Condon principle according to the solid vertical line, the nuclei find themselves at nonequilibrium distance as in part (b), but even more so. Again, vibration ensues at a high level of v'. If the transition terminates at a level close to the energy of dissociation, the atoms may fly apart. In general, an excited state has less room for vibrational levels than does the ground state, as in Figures 4.19 and 5.6, where the potential curve becomes shallow in the excited states. As before, the dashed vertical lines represent less probable transitions, and a v' progression is formed. The band strengths are illustrated in part (c) of Figure 5.7. Since the highest vibrational levels in the excited state now are involved, the band spacings decrease markedly toward higher frequencies. At the dissociation limit the lines, like the vibration levels themselves, merge into an absorption continuum.

Following absorptive transitions to various vibrational levels of an excited state, the molecules can return to lower levels of the ground state by collisional deactivations, by emissive transitions, or by combinations of the two which may be operative at various levels. The main processes usually are competitive. Although collisions are very effective distributors of rotational and vibrational energies, and thereby bring about local thermodynamic equilibrium (LTE), they do not deactivate electronic states very directly. The nature of the difficulty is easily seen. For example, as shown in Figure 4.19 the energy difference between the $^1\Sigma_g^+$ and $^3\Sigma_g^-$ states of the O_2 molecule is about 12,500 cm^{-1} or 5.0×10^{-12} erg. By contrast, the mean energy of molecular translation is given by Eq. (2.41) as 5.7×10^{-14} erg at 0°C, or about 0.01 of the electronic energy. This large disparity implies a low probability that a single collision will remove the electronic energy, which is to say that electronic energy only rarely exchanges with translational energy.

It appears, however, that collisional deactivations of electronic states of some species having complex structures may be accomplished by way of

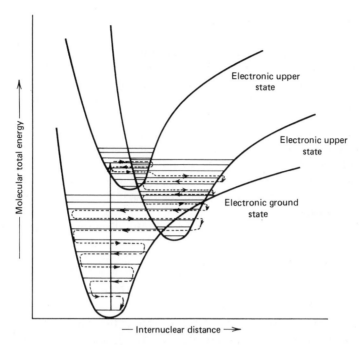

Figure 5.8. Deactivations of electronic upper states by relaxations of vibrational energy and internal conversions to lower states. From Barrow (1964). *The Structure of Molecules.* © 1963 by Benjamin/Cummings Publishing Co., Menlo Park, California.

several vibrational steps and internal conversion of states. This requires that the potential energy curves overlap, as represented in Figure 5.8, where the downward course of energy is shown by the dotted path. Starting with the uppermost state, relaxational collisions carry away the vibrational energy step by step. This continues until the molecule, still in the electronic upper state, reaches a level where its potential energy is approximately equal to that of a higher vibrational level in an electronic intermediate state. If these two states have the same multiplicity, that is, the same number of unpaired electrons, there may be a fair probability that the electronic arrangement will suddenly change to one that is characteristic of an intermediate state. The nuclei continue to vibrate, but now are in a higher level of the intermediate state. That is, at the energy crossover point the molecule has switched its electronic states. Relaxational collisions carry away the energy step by step until the next crossover is reached, and so on. By such means a molecule may go from an upper electronic state to a low vibrational level of the ground state without ever radiating energy.

5.5. ROTATIONAL STRUCTURE OF ELECTRONIC BANDS

The fine structure due to rotational transitions are within the band itself, so that the last two terms of Eq. (5.8) are constant for a given band. Only the rotational term $\Delta\mathscr{E}_{rot}$ needs investigation. If the internuclear equilibrium distance has different values in the two electronic states, the rotational constant (a reciprocal moment of inertia) also has different values. Such differences exist for nearly all electronic states of molecules, and usually the equilibrium distance is greater in the upper states. As an example, in Figure 4.19 they differ by the rather large amount, 0.06 nm, for the $^3\Sigma_u^-$ and $^3\Sigma_g^-$ states of O_2, while they are approximately equal for the $^3\Sigma_u^-$ and $^3\Sigma_u^+$ states. Occasionally, the equilibrium distance in an upper state is less than in a lower state.

Allowing for differences in the equilibrium distances for the upper and lower states, hence differences in the rotational constant, the wavenumber of a rotational line is expressed by

$$\tilde{\nu} = \tilde{\nu}_0 + B_{v'} J'(J' + 1) - B_{v''} J''(J'' + 1) \tag{5.11}$$

where $\tilde{\nu}_0$ is the wavenumber of the vibrational–electronic transition that fixes the band origin. The two rotational terms give the wavenumber difference when the third form of Eq. (4.18) is applied to the levels J' and J'' that are, respectively, in the upper and lower electronic states. The symbols B' and B'' denote the corresponding rotational constants, which usually have different values. As in the earlier treatments of rotational energy, effects of centrifugal stretching and anharmonic vibrations are ignored because they are relatively small. Here they are especially small compared to the effects from differences that may exist between B' and B''.

The upper and lower electronic states may have different values of orbital angular momenta Λ, which is defined by Eq. (4.47) and discussed there. If $\Lambda \neq 0$ for one or both states, the quantum selection rule for J is

$$\Delta J = J' - J'' = 0, \pm 1 \tag{5.12}$$

If $\Lambda = 0$ in both states, which corresponds to a Σ—Σ transition, the rule is

$$\Delta J = J' - J'' = \pm 1 \tag{5.13}$$

The first selection rule produces Q-, R-, and P-branches, while the second one produces only R- and P-branches, just as for a vibration–rotation band. For simplicity in the remainder of this section, the quantum number of the lower state will be denoted by J instead of J''.

Considering first the Q-branch, referred to in Section 5.2 as a *perpendicular branch,* the rule $\Delta J = 0$ means that $J' = J'' = J$. When substituted into Eq. (5.11) this gives

$$Q\text{-branch:}\quad \tilde{\nu} = \tilde{\nu}_0 + J(B_{v'} - B_{v''}) + J^2(B_{v'} - B_{v''}) \qquad (5.14)$$

When the equilibrium distances in the upper and lower states are equal, $B_{v'}$ and $B_{v''}$ are equal and Eq. (5.14) reduces to the wavenumber $\tilde{\nu}_0$ that characterizes the vibrational–electronic transition. This wavenumber marks the band origin, whether the Q-branch is present or not.

For the parallel, low-frequency P-branch, the selection rule is $\Delta J = -1$ so that $J' = J - 1$ where J is used for J''. Hence with this selection rule the upper state quantum number J' is less than the lower state number J. Substituting J for J'' and $J - 1$ for J' in Eq. (5.11) gives for the wavenumbers of the lines,

$$P\text{-branch:}\quad \tilde{\nu} = \tilde{\nu}_0 - J(B_{v'} + B_{v''}) + J^2(B_{v'} - B_{v''}) \qquad (5.15)$$

The wavenumber spacing of lines is obtained by finding the difference of Eq. (5.15) for J and $J - 1$, which is

$$\Delta\tilde{\nu}_P = 2B_{v''} + 2J(B_{v'} - B_{v''}) \qquad (5.16)$$

When the equilibrium distances in the upper and lower states are equal, $B_{v'}$ and $B_{v''}$ are equal, and Eq. (5.16) reduces to

$$\Delta\tilde{\nu}_P = 2B \qquad (5.17)$$

which is the spacing derivable from Eq. (5.2) for adjacent lines in a vibration–rotation band.

For the parallel high-frequency, R-branch, the selection rule is $\Delta J = 1$ so that $J' = J + 1$ where J is used for J''. With this rule the upper state quantum number J' is greater than the lower state number J. By substituting J for J'' and $J + 1$ for J' in Eq. (5.11) we find that the line wavenumbers are

$$R\text{-branch:}\quad \tilde{\nu} = \tilde{\nu}_0 + 2B_{v'} + J(3B_{v'} - B_{v''}) + J^2(B_{v'} - B_{v''}) \qquad (5.18)$$

The wavenumber spacing of lines is obtained by finding the difference of Eq. (5.18) for J and $J + 1$, which is

$$\Delta \bar{\nu}_R = (4B_{v'} - 2B_{v''}) + 2J(B_{v'} - B_{v''}) \qquad (5.19)$$

When the equilibrium distances in the upper and lower states are equal, this reduces to

$$\Delta \bar{\nu}_R = 2B \qquad (5.20)$$

which is the same as Eq. (5.17) for the P-branch.

When the equilibrium distance is greater in the upper state than in the lower, which is frequently true, then $B_{v'} < B_{v''}$ and the difference $B_{v'} - B_{v''}$ has a minus value. For the R-branch this allows the second group of terms in Eq. (5.19) to overtake and exceed the first as J assumes larger values. The line spacing thereby decreases to zero and becomes negative, indicating that the spectral sequence has reversed its direction. The lines of the band thus come to a *band-head* on the larger wavenumber side and then double back on themselves, as shown schematically in Figure 5.9, where the band-head appears at $J' = 4$ in the R-branch. For this condition, that $B_{v'} < B_{v''}$, the line spacing in the P-branch, as given by Eq. (5.16), steadily increases toward smaller wavenumbers as J becomes larger. This is true also for the Q-branch when it is present. The bands are said to be *degraded toward the red*. If $B_{v'} > B_{v''}$, the situation is reversed; the band-head appears in the P-branch; and the bands are *degraded toward the violet*. For all three branches, plots of $\bar{\nu}$ versus J produce parabolic curves because of the terms J^2 in the expressions that define the line wavenumbers. Such plots, known as Fortrat diagrams from their originator, are used in the analysis of band spectra.

5.6. FLUORESCENCE AND PHOSPHORESCENCE

These two related emission phenomena involve one or more excited electronic states and the electronic ground state. Some of the possibilities can be appreciated from Figure 5.10 that shows the vibrational levels of the (singlet) ground state and an excited state, which, for example, may be either $^1\Sigma$ (singlet) or $^3\Pi$ (triplet). The superscript refers to the multiplicity of the state, discussed in Section 4.4.2. Assume now that numerous molecules have been carried by absorptive transitions from the level $v'' = 0$ to several v' levels in the excited state. These transitions correspond to the group of arrows at the left of the figure and follow the Franck–Condon principle. A v' progression in absorption, with $v'' = 0$, thereby is formed in the manner of Table 5.1. If LTE as described in Section 2.4.4 prevails, even approximately, some of these molecules will have been carried to a lower level of the upper state before an emissive transition occurs.

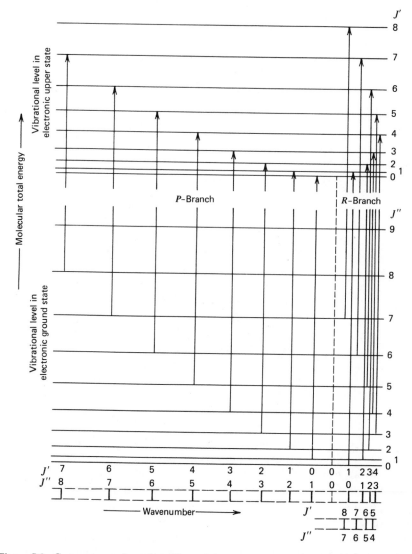

Figure 5.9. Convergence of rotational lines to form a head in an electronic band. The schematic applies to a diatomic species.

Now if the upper state has the same multiplicity as the ground state, a $^1\Sigma$ singlet in this case, an emissive transition to the ground state is not "forbidden." For example, as the figure is drawn the most probable transition from the level $v' = 0$ is to $v'' = 1$, with lesser probabilities for $v'' = 0, 2, 3, \ldots$. This creates a v'' progression in emission, with $v' = 0$. From $v' = 1$ a tran-

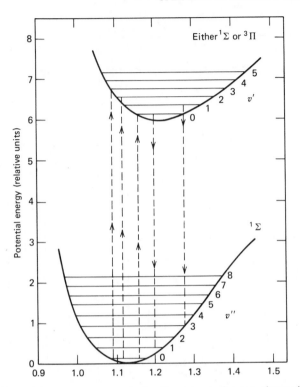

Figure 5.10. Potential energy curves and vibrational levels for two electronic states of a diatomic molecule. Transitions between the v'' and v' levels obey the Franck–Condon principle.

sition to $v'' = 3$ is the most favored while those to $v'' = 1, 2, 4, \ldots$, are less probable. Again a v'' progression is formed but with $v' = 1$. Additional sets of transitions can be deduced from the figure. In each of these cases the emission involves less energy than did the absorption. Hence the wavelengths of the emission lines are greater than those of absorption, and such lines are known as *Stokes lines of fluorescence*.

Considering phosphorescence next, refer to Figure 5.10 and let the excited state now be $^3\Pi$ (triplet). Assume that numerous molecules have been placed in various levels of this state by absorptive transitions from $v'' = 0$. Again, with LTE prevailing many of these molecules will be collisionally deactivated down to the lowest v' levels. Emissive transitions then ensue to several v'' levels, creating progressions as before and likewise involving less energy than did the absorption. However, the energy of the lowest excited triplet state is less than that of the lowest excited singlet state. This may be seen in Figure 4.18 by comparing the Cameron bands with the higher set of

Hopfield–Birge bands. Hence the emission is at still longer wavelengths than the fluorescence. This is *phosphorescence*. Also, the transitions take place between states of different multiplicities, that is, from a triplet to a singlet. These are "forbidden" transitions according to the selection rule (4.57) and the associated explanation. As a result the phosphorescence is generally weak. In contrast, fluorescence is relatively strong because it corresponds to transitions between states of the same multiplicity.

Another important difference between the two phenomena relates to the time scales. Fluorescence usually is emitted within 10^{-6} sec after the exciting light is cut off, and certainly within 10^{-4} sec. Phosphorescence is emitted for times of several seconds after the excitation ceases. With some materials the phosphorescence may persist for many minutes, decaying only slowly. Here we may perceive the action of forbidden (low-probability) transitions; much time must elapse before they are all completed. In effect, the molecules in the lower v' levels of the triplet state are in a metastable condition.

6

Parameters of Line and Band Absorptions

The preceding chapters have emphasized energy levels and transitions, and the organization of lines into bands by the quantum selection rules. But little attention has been given to actual line shapes and strengths, or to their influences on absorption over a band. Here we first examine the standard expressions for Doppler and collisional broadening of lines. The Einstein probability coefficients for transitions are reviewed, and line strengths and absorption coefficients are interpreted in terms of familiar molecular quantities. Finally we look at line and band absorptions as functions of line shape and amount of absorbing gas.

6.1. SHAPES AND WIDTHS OF SPECTRAL LINES

Among the parameters that determine the strength of absorption and emission in a spectral region are line shape and line width. Reviewed summarily in Section 1.4.4, these two parameters are now considered in detail. We begin with natural broadening which, although too small to have practical significance, is useful for explanation. Next, drawing upon the background of kinetic theory in Chapter 2, Doppler broadening and collisional broadening are examined. Finally, attention is called to several correction factors and to the Voigt function for line shape.

6.1.1. Natural Shape and Width

It is convenient to treat natural broadening in terms of emission, but the various factors also apply to absorption. From classical theory, which provides a physical model and leads to about the same result as does quantum theory, an emitting molecule acts like a damped oscillator. The damping represents the energy draining that attends the radiating process and thus limits the life of the oscillation. Such a damped oscillation is not a facsimile of the transition, whose details are masked by the uncertainty principle, but is an analog that enables us to examine the unfamiliar with a familiar light. Garbuny (1965) and especially Stone (1963) should be consulted for fuller treatment than is possible here.

Energy of the damped oscillator as a function of time is expressed by

$$\mathscr{E} = \mathscr{E}_0 \exp(-\gamma t) \tag{6.1}$$

where γ is the natural damping factor. A time constant is defined,

$$\tau_{nat} = \frac{1}{\gamma} \tag{6.2}$$

which is the time required for the oscillator energy to decrease to $1/\exp$ ($=36.8\%$) of the initial value, as by substitution into Eq. (6.1). Usually τ_{nat} is considered the natural lifetime of the excited state. Its value is of the order 10^{-8} sec for the visible region and varies as v^2.

Because the oscillation has a limited duration, its energy is spread across a band of frequencies whose width bears a reciprocal relation to the duration time. The band is centered about the single frequency v_0 that would obtain if the oscillation were undamped, that is, it persisted indefinitely. The frequency spectrum of the oscillation is matched by the spectrum of the radiated energy. At this point we define a spectral emission coefficient $j(v)$, which expresses the energy radiated per unit time, per unit frequency interval, per unit mass of gas, per unit solid angle. A Fourier analysis of the oscillation [see Ditchburn (1977), Levi (1968), or Stone (1963)] yields the Lorentz shape for $j(v)$,

$$j(v) = S \frac{\gamma}{2\pi} \frac{1}{4\pi^2} \frac{1}{(v - v_0)^2 + \gamma^2/4} \tag{6.3}$$

where S is the integrated *line strength*. Very often this quantity is called *line intensity,* but we restrict the latter word to its radiometric meaning of radiant power per unit solid angle. In Eq. (6.3) the term $\gamma/2\pi$ normalizes the expression to meet the requirement

$$S = \int_{-\infty}^{\infty} j(v) \, dv \tag{6.4}$$

Recall now Kirchhoff's law which states that

$$\epsilon_v \equiv \alpha_v$$

Reviewing Eq. (6.3) in this light, we realize that it also defines the mass absorption coefficient $K(v)$ discussed in connection with Eq. (1.30). This means that the shapes of absorption and emission lines are similar if not identical, especially when the optical thickness defined by Eq. (1.33) is small. Therefore, we shall use $j(v)$ and $K(v)$ interchangeably in discussing line shape.

Several parameters of the natural (Lorentz) shape are noted. The maximum value of this distribution is found by solving Eq. (6.3) for the condition

$$j(v) \quad \text{or} \quad K(v) = \frac{2S}{\pi \gamma} \tag{6.5}$$

The half-width α_n, illustrated in Figure 1.10, is that frequency interval between the line center frequency v_0 and the frequency v_h at which $K(v)$ has fallen to one-half of the maximum value. Solving Eq. (6.3) for this condition, we get

$$\alpha_n = \Delta v = v_h - v_0 = \frac{\gamma}{4\pi} = \frac{1}{4\pi \tau_n} \tag{6.6}$$

From Eq. (1.5) we can write this in terms of wavelength,

$$\alpha_n = \Delta\lambda = \frac{\lambda_0^2 \gamma}{4\pi c} \tag{6.7}$$

It can be shown [refer to Garbuny (1965) or White (1934)] that γ varies as $1/\lambda^2$ so that α_n is the same for all wavelengths.

A numerical example is noted. Taking the mercury-198 isotope line at 0.546 μm, discussed following Eq. (1.10), the damping constant $\gamma = 7.35 \times 10^7$. Substitutions into Eq. (6.7) then give

$$\alpha_n \approx 0.58 \times 10^{-5} \text{ nm}$$

Thus the ratio $\lambda/\Delta\lambda = 10^8$, which is far greater than the resolving power of about 10^6 cited in Section 1.2.4 for typical spectrometers. From the viewpoint of quantum mechanics, natural line width is a consequence of the uncertainty principle, Eq. (1.25),

$$\Delta \mathscr{E} \, \Delta t \geq \frac{h}{2\pi} \tag{6.8}$$

In effect, this says that the energy levels, on the time scales of transitions, cannot be considered as infinitely sharp. The levels are diffuse, whereas absolute discreteness would be required for perfectly monochromatic lines.

Until lasers were developed, natural line width represented a mono-chromatic ideal, not attainable in real sources because of Doppler and collision broadening. A gas laser, however, produces line widths several orders of magnitude *less* than natural width, and heterodyne techniques are required for their measurement. Even the early work by Herriot (1962) and Javan et al. (1961) indicated that half-widths of approximately 10^4 Hz were being obtained, and still smaller values for short periods. That they can be achieved at all, in the face of the three broadening processes, is a tribute to the selective amplification that is the basis of lasers.

6.1.2. Doppler Broadening

The molecular velocity components along any direction of observation produce Doppler broadening of spectral lines. This is often called *temperature broadening* because molecular velocity depends on temperature according to Eq. (2.46). Selecting the x axis as a reference, the formula for Doppler shift of frequency due to velocity v_x of either a source or a receiver is

$$\nu' = \nu_0 \left(1 + \frac{v_x}{c}\right) \qquad (6.9)$$

We rearrange it for use here,

$$v_x = \frac{c(\nu' - \nu_0)}{\nu_0} \qquad (6.10)$$

where ν' and ν_0 are the shifted and unshifted frequencies. Each value of v_x produces a corresponding value of ν', a relationship expressed by

$$dv_x = \frac{c}{\nu_0} \, d\nu' \qquad (6.11)$$

That is, to each velocity between v_x and $v_x + dv_x$ there corresponds a frequency between ν' and $\nu' + d\nu'$.

The Maxwell distribution of velocities is given by Eq. (2.51), where the discussion emphasized its probability basis. It is repeated here for convenience,

$$dN = N \left(\frac{m}{2\pi\kappa T}\right)^{1/2} \exp\left(-\frac{m}{2\kappa T} v_x^2\right) dv_x \qquad (6.12)$$

As a reminder, N is the molecular number density and dN is the number of molecules having velocity components between v_x and v_x+dv_x. Each value of v_x produces a particular population dN and a correspondingly shifted frequency v' in Eq. (6.11). Refer now to Eq. (6.4) which relates the emission coefficient $j(v)$ or the absorption coefficient $K(v)$ to the line strength S. The coefficient $K(v)$ at the shifted frequency v' must be proportional to the partial population in the interval dv_x, just as the strength S is proportional to the total population N.

Introducing these ideas into Eq. (6.12) and substituting for v_x and dv_x their equivalents from Eqs. (6.10) and (6.11), we get

$$K(v') = S \frac{c}{v_0} \left(\frac{m}{2\pi\kappa T}\right)^{1/2} \exp\left[-\frac{mc^2}{2\kappa T v_0^2}(v'-v_0)^2\right] dv' \qquad (6.13)$$

This expression is a Gaussian distribution like Eq. (6.12) and is shown as curve (a) in Figure 1.10. The maximum value is given by

$$K(v)_m = S \frac{c}{v_0} \left(\frac{m}{2\pi\kappa T}\right)^{1/2} \qquad (6.14)$$

Noteworthy is the rapid (negative exponential) decrease of $K(v)$ in the wings of the profile. The integral of Eq. (6.13) over all frequencies is equal to S, thus conforming to Eq. (6.4).

The frequency half-width α_D of the Doppler broadened line is found by equating Eq. (6.13) to one-half the value of Eq. (6.14) and solving for $v_h - v_0 \ (=\alpha_D)$. We get

$$\alpha_D = v_h - v_0 = \frac{v_0}{c}\left(\frac{2\kappa T}{m}\ln 2\right)^{1/2} \qquad (6.15)$$

In terms of wavelength this becomes

$$\alpha_D = \lambda_h - \lambda_0 = \frac{\lambda_0}{c}\left(\frac{2\kappa T}{m}\ln 2\right)^{1/2} \qquad (6.16)$$

and in terms of wavenumber,

$$\alpha_D = \tilde{v}_h - \tilde{v}_0 = \tilde{v}_0\left(\frac{2\kappa T}{m}\ln 2\right)^{1/2} \qquad (6.17)$$

directly from Eq. (6.15).

These expressions can be put in a form convenient for calculating by employing the relationships discussed in Section 2.1.3. Multiplying the top and bottom of the parenthetical term by Avogadro's number, Eq. (2.6), and substituting for κ its value in ergs from Eq. (2.47) we obtain for Eq. (6.16),

$$\alpha_D = 3.58 \times 10^{-7} \left(\frac{T}{M_A}\right)^{1/2} \lambda_0 \tag{6.18}$$

and for Eq. (6.17),

$$\alpha_D = 3.58 \times 10^{-7} \left(\frac{T}{M_A}\right)^{1/2} \tilde{\nu}_0 \tag{6.19}$$

where M_A is the molar mass in grams, *numerically* equal to the molecular weight. As an example, we calculate α_D for the rotational lines of CO in the vicinity (low values of J) of the vibrational transition ν_1 which, from Table 4.3, occurs at 4.67 μm. The molecular weight of CO is 28 and a temperature 300 K is assumed. Substitutions into Eq. (6.18) give

$$\alpha_D = 5.5 \times 10^{-3} \text{ nm} \tag{6.20}$$

which is about three orders of magnitude greater than the natural line width given by Eq. (6.8).

6.1.3. Collision Broadening

Radiative transitions are disturbed sufficiently by molecular collisions to cause a broadening of spectral lines that is often called *pressure broadening*. The general role of inelastic collisions in relaxing internal energy to establish local thermodynamic equilibrium (LTE) is reviewed in Section 2.4.4. Here we are concerned with elastic (billiard-ball-type) collisions, which occur much more frequently than inelastic ones. An elastic collision is regarded here simply as a molecular encounter that disturbs, if not interrupts, a radiative transition. After the encounter the transition proceeds to completion, but this latter portion bears only a random phase relation to the original portion. Thus the coherent lifetime is shorter than if undisturbed, and the radiated energy has a greater bandwidth. This is observed as a broadening of the spectral line.

The effective radiating lifetime is determined by the mean free time τ_{coll} between collisions. Assuming that the duration of a collision is so brief that it can be disregarded, the effective lifetime can be taken equal to τ_{coll}, defined by Eq. (2.34) as

$$\tau_{coll} = \frac{\text{mean free path}}{\text{mean speed}} = \frac{\bar{l}}{\bar{v}}$$ (6.21)

Substituting for $\bar{l}$ from Eq. (2.49) and for $\bar{v}$ from Eq. (2.57), we get

$$\tau_{coll} = \frac{(m \kappa T)^{1/2}}{4 \sqrt{\pi}\, d^2\, P} \approx 1.37 \times 10^{-10}\ sec$$ (6.22)

where the numerical value applies to STP conditions. This is about 0.01 of the value cited for τ_{coll} following Eq. (6.2).

The collision alters or destroys the coherency of the wavetrain absorbed or emitted during the transition. A Fourier analysis of the wavetrains yields a frequency spectrum of the energy,

$$K(v) = S \frac{\Gamma}{2\pi} \frac{1}{4\pi^2(v - v_0)^2 + \gamma^2/4}$$ *should be* $\pi^2/4$ (6.23)

This expression, identical in form but not in meaning to Eq. (6.3), also defines a Lorentz shape which is shown in Figure 1.10. It is seen that $K(v)$ decreases slowly, as $(v - v_0)^2$ in the wings of the profile. In Eq. (6.23), S is the line strength according to Eq. (6.4), and Γ is the collision damping constant,

$$\Gamma = \frac{1}{\tau_{coll}}$$ (6.24)

which corresponds to the one defined by Eq. (6.2) for the natural line width. The maximum value of Eq. (6.23) is found to be

$$K(v)_m = \frac{2S}{\pi \Gamma}$$ (6.25)

and the half-width is

$$\alpha_L = \Delta v = (v_h - v_0) = \frac{\Gamma}{4\pi} = \frac{1}{4\pi\tau_{coll}}$$ (6.26)

In terms of wavenumber, Eq. (6.23) is usually written as

$$K(\tilde{\nu}) = \frac{S}{\pi} \frac{\alpha_L}{(\tilde{\nu} - \tilde{\nu}_0)^2 + \alpha_L^2} \qquad (6.27)$$

A similar expression can be written for wavelength.

The effects of temperature and pressure on α_L, are seen from Eqs. (6.22) and (6.26) to be

$$\alpha_L(T, P) = \alpha_{L,0} \frac{P}{P_0} \left(\frac{T_0}{T}\right)^{1/2} \qquad (6.28)$$

where the subscript 0 denotes any reference condition. Equation (6.28) is strictly valid only when the partial pressure of the absorbing gas is a small fraction of the total pressure. This stipulation derives from the fact that collisions between like molecules (self-broadening) produce greater line widths than do collisions between unlike molecules (foreign-gas broadening). Some studies of this effect are noted in the following section. As an example of collision broadening, we find the value of α_L for the CO line whose Doppler width was calculated in Eq. (6.20). Substituting the value of τ_{coll} from Eq. (6.22) into Eq. (6.26) we get for the frequency half-width,

$$\alpha_L = 5.8 \times 10^8 \text{ Hz} \qquad (6.29)$$

In terms of wavelength, using Eq. (1.6) we get

$$\alpha_L = 4.22 \times 10^{-2} \text{ nm} \qquad (6.30)$$

or about eight times the Doppler half-width. In terms of wavenumber the Lorentz half-width is

$$\alpha_L = 1.93 \times 10^{-2} \text{ cm}^{-1} \qquad (6.31)$$

For comparison, the wavenumber spacing $\Delta\tilde{\nu}$ of these lines is 3.86 cm^{-1}, as may be seen from substituting the value of B for the CO entry in Table 4.2 into the relation (4.18c). Hence these lines are well separated even for low values of J.

6.1.4. Line Broadening Considered Further

A principal weakness in simplified calculations of collision broadening based on Eq. (6.22), such as those in the preceding section, lies in the uncertainty of the value of the molecular diameter. A molecule is not really a hard elastic sphere as in the billiard-ball assumption of kinetic theory. The nonsphericity is suggested by the idealized configurations shown in Figure 4.3. Actually, a molecule is surrounded by a force field whose outline and effective range of action are not known precisely. The diameter is not a single absolute value, but depends partly on the phenomenon that is being investigated. For example, the observed half-widths of collision-broadened lines are often two to three times greater than those calculated from the data of kinetic theory. Hence a molecule has an *optical cross section,* which is effective in collision broadening. This is apart from its *kinetic cross section,* which is usually derived from viscosity or heat-conduction experiments.

Also, careful measurements have shown that line shape, in the wings far removed from the line center, do not always correspond to a Lorentz profile. This indicates that collision broadening may involve additional factors such as transition perturbations by the near presence of noncolliding molecules, and the alteration of a radiative to a nonradiative transition when an inelastic collision occurs. Thus a wide range of collisional effects must be allowed. A comprehensive review of these matters is provided by Goody (1964), while Burch et al. (1969) present data on actual deviations from the Lorentz shape.

Line widths also are affected by the distribution of the total molecular population among the various species, both absorbing and nonabsorbing. That is, collisions between like molecules such as CO_2 and CO^2 produce greater broadening than do collisions between unlike molecules such as CO_2 and N_2. In the first case self-broadening and in the second case foreign-gas broadening occur. The second case predominates in the atmospheric infrared because of the relatively low concentrations of the principal absorbers H_2O, CO_2, and O_3. The broadening is expressed by coefficients that give the effect produced by a like gas, and by a specified foreign gas, relative to that produced by N_2. Burch et al. (1962a) have determined self-broadening coefficients for all principal atmospheric species except O_3, and foreign-gas coefficients for several species. Additional investigations of broadening are reported by Anderson et al. (1967), Burch et al. (1969), Chai and Williams (1968), Draegert and Williams (1968), and Tubbs and Williams (1972a, b, c).

Comparisons of Doppler and Lorentz widths, calculated for equal temperatures and pressures, indicate that the Lorentz width is greater by nearly an order of magnitude, as in Eqs. (6.20) and (6.30). As shown in Eq. (6.15), α_D varies as $T^{1/2}$ and has no dependence on P. In the atmosphere, therefore, α_D should decrease slowly to the tropopause in conformance with the lapse rate (2.3b), and then remain nearly constant through the stratosphere where the lapse rate is practically zero. Contrariwise, as shown by Eq. (6.28), α_L varies directly with P and inversely with $T^{1/2}$, hence it decreases more or less exponentially with altitude. Thus Lorentz broadening is more important at lower altitudes, while Doppler broadening predominates in the stratosphere and above.

In the altitude region extending from a lower bound of 20–30 km to an upper bound of 50–70 km, effective line shapes are determined by both broadening processes. Several functions have been devised to express the combined processes in forms suitable for calculation processes. One such is the *Voigt function*, which is a convolution of the Doppler and Lorentz functions according to

$$P(\bar{\nu} - \bar{\nu}_0) \, d\bar{\nu} = \frac{a}{\alpha_D} \frac{\ln 2}{\pi^3} \int_{-\infty}^{\infty} \frac{\exp(-y^2)}{a^2 + (\xi - y)^2} \, dy \, d\bar{\nu} \qquad (6.32)$$

where

$$a = \frac{\alpha_L}{\alpha_D} (\ln 2)^{1/2}$$

$$\xi = \frac{\bar{\nu} - \bar{\nu}_0}{\alpha_D} (\ln 2)^{1/2}$$

$$y = \frac{\bar{\nu}' - \bar{\nu}_0}{\alpha_D}$$

and $P(\bar{\nu} - \bar{\nu}_0)$ is the profile function and $\bar{\nu}'$ is the Lorentz line center for each position on the Doppler-broadened line.

The meaning of the profile function is seen from the relation

$$K(\bar{\nu}) = SP(\bar{\nu} - \bar{\nu}_0) \qquad (6.33)$$

where, as before, $K(\bar{\nu})$ is the absorption coefficient and S is the line strength or integrated absorption coefficient, and the normalization requirement

$$\int_{-\infty}^{\infty} P(\bar{\nu} - \bar{\nu}_0) = 1$$

is met. Information on the Voigt function and examples of its use will be found in Armstrong (1967), Harstad (1972), Kielkopf (1973), Kuhn and London (1969), Posener (1959), and Young (1965). Figure 6.1 shows the Doppler, Lorentz, and Voigt profiles for three different wavenumber regions. In each case the Voigt function reduces the absorption at line center

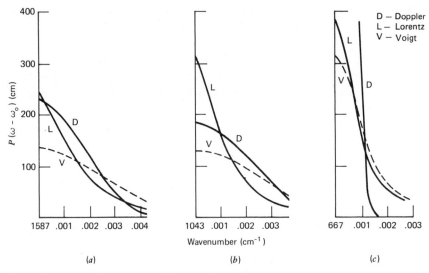

Figure 6.1. Doppler, Lorentz, and Voight profiles corresponding to temperature and pressure at an altitude of 30 km, for infrared lines centered at (a) 1587 cm^{-1} (water vapor), (b) 1043 cm^{-1} (ozone), and (c) 667 cm^{-1} (carbon dioxide). From Kuhn and London (1969).

and moves it into the wings. Although Eq. (6.32) cannot be evaluated in closed form, several analytic approximations and graphical solutions have been devised. These are explained or referenced in the papers cited above. Extensive tutorial treatments are given by Armstrong and Nicholls (1972), Penner (1959), and Tiwari (1978).

6.2. LINE STRENGTHS AND COEFFICIENTS

In this section are presented the molecular and quantum theory factors that govern the strengths of absorption and emission lines. We start with the three Einstein probability coefficients for spontaneous emission, induced emission, and absorption. Several relationships among these coefficients and molecular quantities are developed. This leads to interpretations of line strength and the absorption (emission) coefficient in terms of these quantities.

6.2.1. Transition Probability Coefficients

The strength of a spectral line at a given frequency depends on the population of the initial energy state and the probabilities of three types of transitions: spontaneous emission, induced or stimulated emission (sometimes called negative absorption), and absorption. Their probabilities are governed by factors known as the Einstein (1917) coefficients, summarized here as:

• The coefficient for spontaneous emission, denoted by A_{nm}, gives the probability of a transition, per unit time, from an upper state n to a lower state m. If this lower state is the ground state, A_{nm} is equal to the damping constant γ (reciprocal of the lifetime τ) appearing in Eqs. (6.1) and (6.3). The emitted photon moves away in a random direction, and an equivalent amount of momentum $h\nu/c$, as defined by Eq. (1.20) is imparted to the emitter (atom or molecule) in the reverse direction.

• The coefficient for induced (stimulated) emission B_{nm} gives the probability of a transition from state n to state m, per unit time, per unit radiant energy density in the space surrounding the emitter, per unit frequency interval. The photon is emitted at the same frequency and phase, and in the same direction, as the incident photon. An equivalent amount of momentum is imparted to the emitter in the reverse direction.

• The coefficient for absorption B_{mn} gives the probability of a transition from state m to state n, per unit time, per unit radiant energy density in the space surrounding the emitter, per unit frequency interval. Absorption is an induced process and necessarily takes place in the direction of the incident photon. An equivalent amount of momentum is imparted to the absorber in the forward direction.

The transfer of momentum in each of the processes cited above is a crucial factor in quantum theory. If spherical waves were emitted according to classical theory, momentum would not be exchanged between photon and emitter. However, when the spontaneous transitions are averaged over a large number of events, they converge to the classical picture, leaving unaltered the net momentum of the emitter. A series of absorptions (a stimulated process) and stimulated emissions following each other closely is the basis, of course, for gas-laser operation, which also leaves unaltered the net momenta of the emitters.

Relationships among the coefficients and molecular quantities are now derived, following, generally, the treatments by Bauman (1962), Ditchburn (1958, 1977), Garbuny (1965), and Herzberg (1961). Considering a gas that is in radiative equilibrium, the sum of the rates for spontaneous and induced emissions must equal the rate for absorptions. That is,

$$N_m B_{mn} w(\nu)\, dt = N_n \left[B_{nm} w(\nu) + A_{nm} \right] dt \qquad (6.34)$$

where N_n and N_m are the populations of the upper and lower states, respectively, and $w(\nu)$ is the radiant density defined in Appendix C. We assume for simplicity that the energy levels are not degenerate, thus their statistical weights are equal to unity. The populations of the levels then are related by Eq. (2.61), which in present terms is written as

$$N_n = N_m \exp - \frac{\mathscr{E}_n - \mathscr{E}_m}{\kappa T} \tag{6.35}$$

This allows Eq. (6.34) to be rewritten, after some manipulations, in the form

$$w(\bar{\nu}) = \frac{A_{nm}}{B_{nm}} \frac{1}{(B_{mn}/B_{nm}) \exp(h\nu/\kappa T) - 1} \tag{6.36}$$

which must agree with the Planck law (1.46) for the radiant energy density of a volume of space. The law is

$$w(\nu) = \frac{8\pi h \nu^3}{c^3} \times \frac{1}{\exp(h\nu/\kappa T) - 1} \quad \text{J cm}^{-3} \text{ Hz}^{-1} \tag{6.37}$$

Such agreement between Eqs. (6.36) and (6.37) requires

$$B_{mn} \equiv B_{nm} \tag{6.38}$$

and

$$A_{nm} \equiv \frac{8\pi h \nu^3}{c^3} B_{nm} \tag{6.39}$$

Several comments should be made at this point. When radiative equilibrium exists, the transfers of momenta between the molecules and the photons of the radiation environment do not disturb the Maxwell distribution (2.54) of molecular speeds. In fact, if the collision rate is less than the photon absorption–emission rate, the speed distribution may well be established by such transfers. The equality of B_{mn} and B_{nm} means that a photon, in encountering an energy level for which the transition energies to a higher and to a lower level are each equal to the photon energy, has equal probabilities of inducing absorption and of inducing emission. Because the populations of the lower levels normally are greater than those of upper levels according to Eq. (6.35), absorptive transitions are more numerous than induced emissive transitions. However, the radiative account is balanced, that is, equilibrium is maintained, by the spontaneous emissive transitions according to Eq. (6.34). Induced emission in the atmosphere usually may be ignored.

6.2.2. Relationships of Einstein Coefficients

At this point, it is instructive to define the three Einstein coefficients in terms of molecular quantities, since Eq. (6.39) defines only a ratio. Invoking classical theory, the average rate of energy emission into 4π sr by a molecular dipole, in Gaussian units, is (see Frank, 1950)

$$\frac{d\mathscr{E}}{dt} = \Phi = \frac{\omega^4 \mu_0^2}{3c^3} = \frac{16\pi^4 \nu^4 \mu_0^2}{3c^3} \tag{6.40}$$

where Φ is the radiant power as defined in Appendix C. The factor μ_0 is the dipole moment, defined by Eq. (4.2) as a molecular quantity equal to the product of discrete charge and the distance between effective centers of positive and negative charge. In quantum mechanics, however, the exact positions or orbits of electrons in atoms and molecules cannot be known. Instead, it is necessary to treat the group of charges as a continuous distribution, where the region of greatest probability density is expressed by matrix elements derived from the wave equations. Also, during the transition, the dipole moment is a mixture of the initial and final levels. The moment in matrix form is defined by

$$R_{nm} = Q \int \psi_n^* d \psi_m dV \tag{6.41}$$

where d is the effective separation of the unlike charges, ψ_n and ψ_m are the wave functions of the upper and lower levels, the asterisk denotes the complex conjugate, Q is the charge, and dV is the volume element. The subscripts of R can be in either sequence.

As shown by Garbuny (1965) and indirectly by Eisberg and Resnick (1974), the amplitude d_0 of the dipole oscillation is related to the matrix element by

$$d_0^2 = 4 \left| R_{nm} \right|^2 \tag{6.42}$$

Thus from Eq. (4.2) the square of the dipole moment becomes

$$\mu_0^2 = 4Q^2 \left| R_{nm} \right|^2 \tag{6.43}$$

Substituting this into Eq. (6.40) we obtain for the average rate of energy emission by the dipole, that is, the power emitted,

$$\overline{\frac{d\mathscr{E}}{dt}} = \Phi = \frac{64\pi^4\nu^4Q^2|R_{nm}|^2}{3c^3} \tag{6.44}$$

Since this energy is carried away by photons of energy value $\mathscr{E} = h\nu$, the time rate Φ' of photon emission is

$$\Phi' = \frac{\Phi}{h\nu} = \frac{64\pi^4\nu^3Q^2|R_{nm}|^2}{3hc^3} \tag{6.45}$$

Returning now to the Einstein coefficient A_{nm} for spontaneous emission, the rate of photon emission from level n to level m per molecule is equal to the value of that coefficient. This assumes that the molecule is returned to level n, either by absorptive or collisional transition, immediately after its downward transition to level m. Hence we can equate A_{nm} to Eq. (6.45),

$$A_{nm} = \frac{64\pi^4\nu^3Q^2|R_{nm}|^2}{3hc^3} \tag{6.46}$$

which defines, in a mixed manner, the Einstein coefficient for spontaneous emission in terms of familiar quantities. In cgs units the dimensions of charge Q are $L^{3/2}M^{1/2}T^{-1}$ and the dimension of R_{nm} is L. The dimension of A_{nm}, therefore, is T^{-1}, which is implied in its definition at the beginning of Section 6.2.1.

Order-of-magnitude values for A_{nm} are 10^8, 10, and 1 sec^{-1} for electronic, vibrational, and rotational transitions, respectively. When the emissive transition is to the ground state, we have

$$A_{nm} = \gamma$$

as noted earlier. Hence the average lifetimes of excited states are approximately equal to the reciprocals of the above values. This means that the vibrational and rotational lifetimes in the lower atmosphere are far greater than the mean time between molecular collisions, found from Eq. (2.34) as about 10^{-10} sec. Consequently, there is ample opportunity for collisional relaxation of internal energy and prevalence of local thermodynamic equilibrium (LTE).

Considering next the coefficient B_{nm} for induced emission and B_{mn} for absorption, the identity (6.38) requires that the statistical weights g of levels m and n be equal. The condition

$$g_n B_{nm} = g_m B_{mn}$$

must obtain. If the levels are nondegenerate, then the statistical weights are each equal to unity. For such cases we combine Eqs. (6.38) and (6.39) with Eq. (6.46) to obtain

$$B_{nm} = B_{mn} = \frac{8\pi^3 Q^2 |R_{nm}|^2}{3h^2} \qquad (6.47)$$

which has the dimension $M^{-1}L$.

6.2.3. Line Strength and Absorption Coefficient

The strength of an emission or absorption line can now be expressed in terms of molecular quantities. The total power at frequency v, per unit frequency interval, emitted into 4π sr, by a unit volume of gas containing N_n emitters maintained by LTE in the nth energy level, is

$$\Phi_{em} = N_n A_{nm} h v_{nm} \qquad (6.48)$$

Substituting for A_{nm} from Eq. (6.46) we obtain

$$\Phi_{em} = \frac{64\pi^4 v_{nm}^4 Q^2 |R_{nm}|^2}{3c^3} N_m \qquad (6.49)$$

where the effects of self-absorption are ignored. This power is emitted isotropically because of the random orientations of the emitters and the random times of the emissions. The dependence on v^4 should be noted.

The strength of absorption is difficult to define simply because of its rapid variation with frequency, as in Figure 6.1. When the optical thickness is small, however, these effects can be disregarded. The power absorbed per unit volume, at frequency v and per unit frequency interval, is, from Eq. (6.34),

$$\Phi_{ab} = N_m B_{mn} h v_{mn} w(v) \qquad (6.50)$$

for N_m absorbers in the mth energy level, for radiant energy density $w(v)$. Substituting for B_{mn} from Eq. (6.47) we get

$$\Phi_{ab} = \frac{8\pi^3 v_{mn} Q^2 |R_{mn}|^2}{3h} N_m w(v) \qquad (6.51)$$

Absorption strength can also be stated in terms of the absorption coefficient applicable to the attenuation along an optical path. This coefficient was discussed in Section 1.4.3, and now we can relate it to the molecular quantities. Consider first that the radiant power per unit cross section of an optical path is defined by the radiometric quantity *irradiance E*, which is equal to the product of the radiant energy density and the velocity of light. Hence Eq. (6.50) can be written

$$\Phi_{ab} = \frac{N_m B_{mn} h \nu_{mn}}{c} E \qquad (6.52)$$

Because the absorbed power represents the irradiance decrease dE across a path lamina dx, we have for the fractional decrease,

$$\frac{dE}{E} = \frac{-N_m B_{mn} h \nu_{mn}}{c} dx \qquad (6.53)$$

A comparison of this expression with Eq. (1.28) shows that the two have the same form; that is, the fractional loss of either irradiance or radiance (if the beam solid angle must be taken into account) is equal to the product of an energy-absorbing term per unit of path length (of unit cross section) and an infinitesimal length of path.

Thus we can write

$$k(\nu) = \frac{B_{mn} h \nu_{mn}}{c} N_m \qquad (6.54)$$

or in molecular terms, by substituting for B_{mn} from Eq. (6.47),

$$k(\nu) = \frac{8\pi^3 \nu_{mn} Q^2 |R_{mn}|^2}{3hc} N_m \qquad (6.55)$$

for the absorption coefficient applicable to unit path length, and having the dimension L^{-1}. Despite this dimension, Eq. (6.55) defines a volume coefficient in that the number N_m of molecules in energy level m per unit volume specifies the amount of absorber. If no foreign gas were present, the ratio of N_m to the number density N would depend only on temperature and the statistical weights of the energy levels, according to Eq. (3.75). But foreign gases are always present in the atmosphere, and the number density itself is a function of temperature and pressure. A more definite form of coefficient is obtained by adjusting the value of $k(\nu)$ so that it refers to a unit volume of the selected species at STP conditions, or to unit mass of that species. Unit cross section of the path is understood. Sometimes the coefficient is stated in terms of the *atm cm,* discussed in connection with Table 1.1. When the gas is other than water vapor, an atm cm is equivalent to a 1 cm^3 volume.

The absorption also can be referred to a single molecule, that is, to the molecular cross section, that by absorption is effective in removing radiant flux from the beam. Such an absorption cross section corresponds to a scattering cross section. When $k(v)$ refers to a unit volume of STP gas, the absorption cross section σ of a molecule is defined by

$$\sigma(v) = \frac{k(v)}{N_L} \qquad (6.56)$$

where N_L is Loschmidt's number, Eq. (2.7). The dimension of $\sigma(v)$ is L^2, that of an area. This cross section is extensively used in studies of ultraviolet absorption.

Another version of the coefficient refers to unit mass of the gas per unit cross section, as noted in Eq. (1.30). The dimensions of this mass coefficient $k(v)$ are $M^{-1}L^2$, and it is expressed as a numeric per g per cm^2. The amount of absorber along the path then is stated in units of g cm^{-2}. In the case of water vapor, 1 g cm^{-2} is equal to 1 pw cm.

6.3. ABSORPTION BY LINES AND BANDS

Although absorption and emission usually are observed across a band, whose width is set by measurement objectives and instrument capabilities, there is great interest in the spectral line itself as an observable. This interest has been increased, of course, by the widespread use of lasers. Beyond this, spectral lines form the structure of band absorption and emission and are important for this reason also. Here we look first at the ways in which the absorptions by lines are governed by their widths and the amount of the absorber that creates the line. This leads to a review of the principal models that have been devised to calculate the absorptions by bands of many lines. The behavior of the Elsasser and statistical models is then discussed in terms of temperature, pressure, and amount of absorber. Attention is given to the *weak-line* and *strong-line* approximations, which simplify the problems of calculation.

The subject of this section is complex, and we can do no more than present the salient features. We wish, however, to direct readers to authoritative treatments which give important details. Herzberg (1961) and Walker and Straw (1967) are, as usual, good overall references, although their emphasis is on spectra rather than absorption models. Herzberg (1966) contains a vast amount of information on electronic spectra, but is not for the novice. Herzberg (1971) is a remarkably concise and valuable reference on spectroscopy. Goody (1964) is a standard reference on line and band models, while Zuev (1974) gives an excellent treatment with much applicational data. Shorter coverage, but adequate for many purposes, is provided by Jamieson et al. (1963) and Kondratyev (1969). Detailed attention to certain aspects, indicated by the titles, will be found in Goldman and Kyle (1968) and Ludwig et al. (1973). Various band models are applied to the calculation of transmission functions by La Rocca and Turner (1975). Tiwari (1978) provides a thorough review of all band models and line profiles, with emphasis on simple analytical expressions. Armstrong and Nicholls (1972) emphasize the physics in their complete tutorial presentation of line and band absorptions and emissions.

6.3.1. Line Absorption

We consider an isolated line that is entirely within a wavenumber interval $\Delta \bar{\nu}$. The mean fractional absorptance $\overline{A}$ over the interval is, from Lambert's law, Eqs. (1.30b) and (D.4),

$$\overline{A} = \frac{1}{\Delta \bar{\nu}} \int_{\Delta \bar{\nu}} \{1 - \exp[-K(\bar{\nu})u]\} d\bar{\nu} \qquad (6.57)$$

where $K(\bar{\nu})$ is the mass absorption coefficient and u is the mass of absorber per unit cross section of path. Employing the Lorentz line for an example, substituting for $K(\bar{\nu})$ from Eq. (6.27) gives

$$\overline{A} = \frac{1}{\Delta \bar{\nu}} \int_{\Delta \bar{\nu}} \left\{1 - \exp\left[-\frac{S\alpha_L u}{\pi(\bar{\nu} - \bar{\nu}_0)^2 + \alpha_L^2}\right]\right\} d\bar{\nu} \qquad (6.58)$$

This is related to the *equivalent width* $W(u)$ of a line by

$$W(u) = \overline{A} \Delta \bar{\nu} = \int_{\Delta \bar{\nu}} \left\{1 - \exp\left[-\frac{S\alpha_L u}{\pi(\bar{\nu} - \bar{\nu}_0)^2 + \alpha_L^2}\right]\right\} d\bar{\nu} \qquad (6.59)$$

This name refers to the width of a totally absorbing line having the same integrated absorptance as the given line. Such a line would have a rectangular profile. The dependence $W(u)$ expressed by Eq. (6.59) is called the "curve of growth."

An exact solution of Eq. (6.59) was found by Ladenberg and Reiche (1913) to be

$$\overline{A} = 2\pi\alpha_L x \exp(-x)[I_0(x) + I_1(x)] \qquad (6.60)$$

where

$$x = \frac{Su}{2\pi\alpha_L} \qquad (6.61)$$

and $I_0(x)$ and $I_1(x)$ are the Bessel functions of order zero and one, of imaginary argument. The Ladenberg and Reiche function

$$f(x) = x \exp(-x)[I_0(x) + I_1(x)] \qquad (6.62)$$

has been computed by Kaplan and Eggers (1956) and values are tabulated in Goody (1964). Calculations for lines with combined Doppler and Lorentz profiles have been made by Yamada (1968), and for Voigt profiles by Jansson and Korb (1968) and Kyle (1968).

When x is much less than unity, an approximate solution of Eq. (6.60) is

$$\overline{A} = \frac{2\pi\alpha_L x}{\Delta\bar{\nu}} = \frac{Su}{\Delta\bar{\nu}} \qquad (6.63)$$

This is known as the *weak-line approximation,* which, according to Plass (1958) is accurate within 10% when $x < 0.2$. In this case the absorptance varies linearly with the integrated line strength S, and linearly with the amount u of absorber. The absorptance is small even at line center, and necessarily is much smaller in the wings, as shown by curve (a) of Figure 6.2. Because the absorptance is small across the entire line, all of the molecules along the path are exposed to nearly equal incident fluxes. Stated differently, the molecules at the far end of the path are not "shielded" by those at the near end. This is the reason that the absorptance varies linearly with the total amount of absorber, that is, with the total number of molecules.

When x is much greater than unity, an approximate solution of Eq. (6.60) is

$$\overline{A} = \frac{2}{\Delta\bar{\nu}}(S\alpha_L u)^{1/2} \qquad (6.64)$$

This is the *strong-line approximation,* accurate within 10% when $x > 1.63$, according to Plass (1958). The absorptance now varies as the square roots of line strength, line width, and amount of absorber. Because of the dependence on line width, the absorptance varies as the square root of pres-

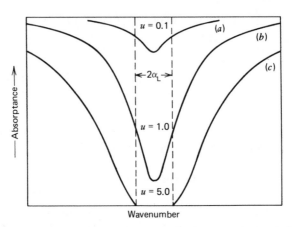

Figure 6.2. Absorptance of single spectral line. Curve (a) refers to the *weak-line* approximation (linear regime). Curves (b) and (c) refer to the *strong-line* approximation (square-root regime). From Elsasser and Culbertson (1960).

sure and inversely as the fourth root of temperature, as may be understood from Eq. (6.28). Such weak dependence on temperature is usually ignored in practice. Absorptance according to the strong-line approximation is shown progressively by curves (b) and (c) of Figure 6.2. As the amount u of absorber is increased, the absorptance in the central region of the line reaches its maximum value of unity. The center of the line is "eaten away." When the amount of absorber is increased still further, additional absorption can occur only in the wings, and the overall absorptance increases only slowly with amount of absorber.

Pressure and amount of absorber along an optical path in the atmosphere vary over extremely wide ranges. Generalized relationships between these two parameters and the absorption process are shown in Figure 6.3. Since these are log-log plots, the slope of the curve at any point is equal to the exponent of the variable governing the absorption. For example, the curves tend to a slope of unity for $u \ll 1$. This is the condition of linear absorption that is defined by Eq. (6.63) and illustrated on a spectral basis by curve (a) of Figure 6.2. When $u \gg 1$, the slopes in Figure 6.3 become approximately one-half, signifying that the absorption varies with $u^{1/2}$ as in Eq. (6.64). For this condition the absorption, already complete in the central region of the line, increases only in the wings far beyond the half-width points, as in curve (c) of Figure 6.2.

When the Doppler shape defined by Eq. (6.13) predominates, which is the case at altitudes above 40 km or so, that expression is substituted into Eq. (6.57) after replacing v by $\bar{v}$. Although the resulting integral is not in closed form, Plass and Fivel (1953) developed approximate solutions by

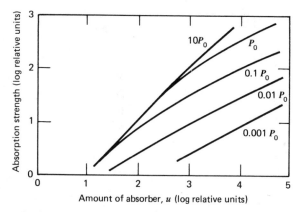

Figure 6.3. Strength of absorption by a Lorentz line as a function of pressure and amount of absorber. The indicated pressures are multiples of some standard pressure P_0. From Jamieson et al., *Infrared Physics and Engineering.* © 1963 by McGraw-Hill, Inc. Used with the permission of McGraw-Hill Book Co.

employing the techniques that yielded Eqs. (6.63) and (6.64). When x as defined by Eq. (6.61) is less than unity, they found that

$$\overline{A} = \frac{Su}{\Delta \bar{\nu}} \tag{6.65}$$

which is the same as Eq. (6.63) for the Lorentz shape. Thus when the absorptance is small, the exact line shape is not important because every molecule, even at the far end of the path, is absorbing practically its full share of the incident flux. For large values of the absorptance, they defined a parameter x_D

$$x_D = \left(\frac{\ln 2}{\pi}\right)^{1/2} \frac{Su}{\alpha_D} \tag{6.66}$$

and found that the absorptance could be expressed by

$$\overline{A} = \frac{1}{\Delta \bar{\nu}} \, 2\alpha_D \left(\frac{\ln x_D}{\ln 2}\right)^{1/2} \tag{6.67}$$

where α_D is the wavenumber half-width, Eq. (6.17).

The absorption process by a Doppler-broadened line, as a function of u, for three temperatures is shown in Figure 6.4. At small values of u, the absorption increases linearly with u according to Eq. (6.65) and is the same for all temperatures. When u becomes large, the absorption increases only

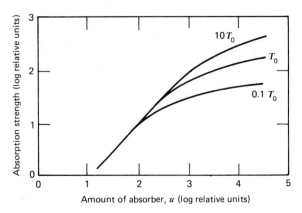

Figure 6.4. Strength of absorption by Doppler line as a function of temperature and amount of absorber. The indicated temperatures are multiples of some standard temperature T_0. From Jamieson et al., *Infrared Physics and Engineering.* © 1963 by McGraw-Hill, Inc. Used with the permission of McGraw-Hill Book Co.

as $(\ln u)^{1/2}$, as required by Eqs. (6.66) and (6.67) and indicated by the flattening of the curves. This is not a strong dependence and means that path lengths at high altitudes can be increased considerably without greatly decreasing the transmittance. When combined Doppler and pressure broadening must be considered, the difference between the two line shapes illustrated in Figures 1.10 and 6.1 becomes important. The Doppler shape decays rapidly as

$$\exp[-(\tilde{\nu} - \tilde{\nu}_0)^2 \times \text{constant}]$$

according to Eq. (6.13). The Lorentz shape decays slowly at frequencies well removed from line center according to

$$[(\tilde{\nu} - \tilde{\nu}_0)^2 + \text{constant}]^{-1}$$

as in Eq. (6.27). Thus the Lorentz shape, even at small half-widths, may produce greater absorption in the far wings than does the Doppler shape.

6.3.2. Types of Band Absorption Models

Although the spectral line absorptions exhibit well-defined relationships to the physical parameters, the absorptions over spectral bands are more difficult to describe analytically. A complete vibration–rotation band, or even a narrower band of interest, may contain hundreds of lines, particularly when vibrational combination frequencies are present. The lines have

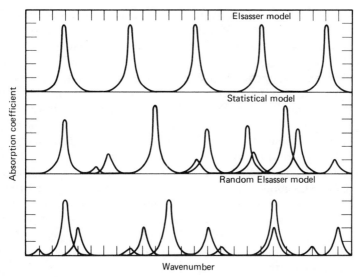

Figure 6.5. Idealized representations of three band absorption models. From Jamieson et al. (1963) *Infrared Physics and Engineering.* © 1963 by McGraw-Hill, Inc. Used with the permission of McGraw-Hill Book Co.

varying strengths, and many of them overlap in their wings, resulting in less absorptance than if the lines were isolated. The absorptance of such a band can be calculated by defining all the parameters of each line and then summing the contributions of all the lines. Such a procedure is known as direct integration across the band, and its accuracy obviously depends on a detailed knowledge of band structure. Calculations of this kind have been made by Gates and Harrop (1963) and Gates et al. (1963, 1964) for the 2.7-μm H_2O band, by Drayson (1965, 1966) and Drayson et al. (1967, 1968) for the 15-μm CO_2 band, and by Shaw and Houghton (1964) for the 4.7-μm CO band. Such calculations were almost prohibitive before the availability of electronic digital computers.

Calculations of band absorptance are greatly eased by employing a band model where the line strengths and spacings follow certain selected mathematical rules. Several such models have been devised to represent band absorptions in various regions of the spectrum with reasonable accuracy. Those most frequently encountered are the Elsasser or regular model, the Mayer–Goody statistical model, the random Elsasser model, and the quasirandom model. Schematic arrangements of the lines in the first three models are shown in Figure 6.5. In the regular model, described by Elsasser (1938, 1942) and Elsasser and Culbertson (1960), both the line strengths and spacings are unchanging across the band, that is, the line is allowed to repeat itself, as seen in the top panel of the figure. In the statistical model devised by Goody (1952, 1964), the lines have random spacings and their strengths can vary according to a selected distribution function. Only these two models are discussed in this book. In general, any band model is strictly valid only for a homogeneous path at constant temperature.

Additional band models are not lacking, however. In the random Elsasser model (see Plass, 1958 or Kaplan, 1954) regular Elsasser bands having different line spacings and strengths are superposed, as shown in the bottom panel of Figure 6.5. In the quasirandom model devised by Stull et al. (1962), the absorptances are computed for randomly placed lines in narrow intervals, and these are averaged to obtain the overall absorptance of the band. This model appears to be the most accurate of the four, but also requires the most computation. Wyatt et al. (1964) and Stull et al. (1964) have used this model for their extensive tables of H_2O and CO_2 transmittances described in Sections 7.1.2 and 7.2.2. King (1964, 1965) has developed a generalized model that mediates the Elsasser and statistical models, and McClatchey et al. (1972) have employed this model in constructing a very useful set of transmittance charts.

6.3.3. Regular Elsasser Model

The lines in a regular Elsasser model have constant strength, half-width, and spacing across the band. This model is well adapted to the CO_2 bands which have fairly regular rotational structures. It is less well adapted to the irregular structures of the H_2O and O_3 bands. Assuming a Lorentz line shape defined by Eq. (6.27) in terms of wavenumber, the spectral absorption coefficient for a band having an infinite array of identical lines, at wavenumber $\tilde{\nu}$ from any line center, is

$$K(\tilde{\nu}) = \sum_{n=-\infty}^{n=\infty} \frac{S\alpha_L/\pi}{(\tilde{\nu} - nd)^2 + \alpha_L^2} \qquad (6.68)$$

where S is the line strength defined by Eq. (6.4) and d is the wavenumber distance between adjacent line centers. The mean absorptance of such a band (see Elsasser, 1938 or Jamieson et al., 1963) is

$$\overline{A} = 1 - \frac{1}{2\pi} \int_{-\pi}^{\pi} \exp\left(-\frac{\beta x \, \sinh\beta}{\cosh \beta - \cos s}\right) ds \qquad (6.69)$$

where

$$\beta = \frac{2\pi\alpha_L}{d}$$

$$s = \frac{2\pi\tilde{\nu}}{d}$$

and x is defined by Eq. (6.61). Seitz and Lundholm (1964) have developed a series that represents the integral in Eq. (6.69) for all parameter values. Values of the integral for $\beta \leq 1$ have been tabulated by Kaplan (1954) and by Wark and Wolk (1960).

Two limiting cases of Eq. (6.69), known as the *weak-* and *strong-line approximations*, may be expressed in simple terms. When the absorption is weak at all wavenumbers in the band, the absorption is a function of only

$$\beta x = \frac{Su}{d}$$

and the weak-line approximation is given by

$$\overline{A} = 1 - \exp(-\beta x) \tag{6.70}$$

which is equivalent to Bouguer's law, Eq. (1.29). This means that even a complex band absorbs according to a simple law when the absorption near the centers of even the strongest lines is weak. The particular arrangement of the lines in the band then does not matter because the absorptions of the separate lines are additive, and the total effect increases linearly with the amount of absorbing material. Equation (6.70) gives results that are within 10% of the exact results from Eq. (6.69) when $x < 0.2$ and $\beta < 0.2$, or for any value of x when $\beta > 1$.

The strong-line limit of Eq. (6.69) occurs when the absorptions are sufficiently strong that the centers of the lines are opaque (non-transmitting), a condition represented by curve (c) of Figure 6.2. The absorption now is a function of only

$$\beta^2 x = 2\pi\alpha_L S\mu/d^2$$

and the strong-line approximation is given by

$$\overline{A} = \text{erf} \left(\frac{\beta^2}{2}x\right)^{1/2} \tag{6.71}$$

where erf(z) is the error function integral defined by

$$\text{erf}(z) = \frac{2}{\pi^{1/2}} \int_0^z \exp(-z^2)dz \tag{6.72}$$

According to Plass (1960), the approximation (6.71) gives results that are always within 10% of the exact results obtainable from Eq. (6.69) when x > 1.63. Equation (6.71) is the basis for the tables of H_2O and CO_2 transmittances computed by Passman and Larmore (1956). These tables are reproduced in Hudson (1969), with extensions to longer wavelengths by that author.

6.3.4.   Statistical or Random Model

For the statistical model, consider a spectral interval $\Delta\bar{\nu}$ such that

$$\Delta\bar{\nu} = nd$$

where d is the mean distance between the centers of the n randomly appearing lines. Let $\mathcal{P}(S_i)$ be the probability that the ith line has a strength S_i, with $\mathcal{P}$ normalized so that

$$\int_0^\infty \mathcal{P}(S)\, dS = 1 \tag{6.73}$$

A general expression for the absorptance over the interval $\Delta\bar{\nu}$ is

$$\overline{A} = 1 - \exp\left[-\frac{1}{d} \int_{\Delta\bar{\nu}} \mathcal{P}(S)(A_i\Delta\bar{\nu})dS \right] \tag{6.74}$$

when the number of lines is large. Here A_i is the absorptance of a single line over $\Delta\bar{\nu}$, and

$$(A_i\Delta\bar{\nu}) = \int_{\Delta\bar{\nu}} \{1 - \exp[-K(\bar{\nu})u]\}\, dv \tag{6.75}$$

The expression (6.74) is valid for any line shape according to Elsasser and Culbertson (1960).

We next have to consider the weak- and strong-line approximations for each of two cases: (1) the line strengths are equal, and (2) the line strengths can be represented by a simple function, for example, the Poisson (essentially exponential) distribution. The approximations are given by Plass

(1960) and Wolfe (1965). For *equal line strengths* the integral in Eq. (6.74) can be omitted, and the weak-line approximation is

$$\overline{A} = 1 - \exp(-nA_i) \tag{6.76}$$

which is equivalent to Eq. (6.70) for the weak-line approximation of the Elsasser model. When A_i is small, meaning that the lines are absorbing very weakly, the series expansion of the exponential reduces Eq. (6.76) to the sum of the individual line absorptions, or the product nA_i. The strong-line approximation is

$$\overline{A} = 1 - \exp\left(-4 \, \frac{Su\alpha_L}{d^2}\right)^{1/2} \tag{6.77}$$

When the line strengths follow a *Poisson distribution,* namely,

$$\mathscr{P}(S) = S_0 \exp\left(-\frac{S}{S_0}\right) \tag{6.78}$$

where $\mathscr{P}$ is a normalized probability, and S_0 is a mean line strength, the weak-line limit is

$$\overline{A} = 1 - \exp(-\beta x) \tag{6.79}$$

which is the same as Eq. (6.70) for the weak-line limit of the Elsasser model. The strong-line approximation, for the frequent atmospheric case in which the line half-widths are less than the line spacings, is

$$\overline{A} = 1 - \exp\left(-\frac{S_0 u \pi \alpha_L}{d}\right) \tag{6.80}$$

This differs from the corresponding expression Eq. (6.77) for lines of equal strength only by a numerical factor in the exponential and the presence of S_0 instead of S.

7

Absorption and Emission Data

The data of absorption and emission by atmospheric gases are the links between the real world and the theories and principles discussed in previous chapters. There is no lack of data for consideration; the problems of a one-chapter treatment are those of selection. Here we single out for discussion those sets of measurements that, as reported by the investigators, do illustrate the principles set forth in previous chapters. For each of the principal absorbing gases additional measurements are cited to apprise readers of the large amounts of data that are available.

7.1. WATER VAPOR ABSORPTION

Because of its importance to the earth's heat balance and the operation of infrared equipment, the absorption by water vapor has been investigated thoroughly. In this section we review the key molecular factors for vibration and rotation and relate the vibrational modes to the observed absorption bands. Attention is then directed to the numerous sets of absorption parameters and transmittance values that have been compiled. Representative absorption data are then reviewed and several measurement programs are referenced.

7.1.1. Molecular Factors

The H_2O molecule has an asymmetric top (bent triatomic) configuration, as depicted in Figure 4.3, and the three rotational constants are listed in Table 4.2. Using these values and employing the selection rule $\Delta J = \pm 1$, the pure rotational lines can be computed from Eq. (4.18), according to the discussion in Section 4.2.3. Because these rotational constants have relatively large values and differ considerably among themselves, the spacings of the lines for the three degrees of rotational freedom tend to be wide and irregular. The resulting spectrum has a ragged appearance. Complexities are introduced by other isotopic forms of water vapor present in fractional percentages. These forms produce weak but identifiable rotational lines, differing slightly in frequency from those of $HH^{16}O$. The complete rotation spectrum, or rotation band, of water vapor extends from about 15 μm to millimeter wavelengths. Through most of this extreme-infrared region the absorption lines are sufficiently numerous and overlapping to render the sea-level atmosphere opaque for distances greater than several meters. Populations of the rotational levels conform to Eqs. (4.22) and (4.28).

The vibrational modes and parameters of the H_2O molecule are listed in Table 7.1, which combines information from Table 4.3 and Figure 4.14. A transition at each of the fundamental frequencies gives rise to a vibration–rotation band, as illustrated schematically by Figure 5.2. At usual atmospheric temperatures most of the molecular population is in the vibrational level $v = 0$, as will be realized from Eq. (4.42) and Figure 4.13. When transitions occur between *nonadjacent* levels, weaker overtone frequencies are produced. Generally, these are not exact integer multiples of the fundamental when the vibration is anharmonic, as discussed in Section 4.3.2. This also permits simultaneous transitions in two of the modes, creating weak combination and difference frequencies. The fundamental band at v_2 has the greatest atmospheric effect because of its strength and width, which overlaps portions of both the solar and terrestrial spectrum. The bands at v_1 and v_3 overlap due to the small difference of their center frequencies. These two bands, along with the overtone band at $2v_2$ centered at 3.17 μm, are often lumped together in dealing with low-resolution absorption in this part of the spectrum. Absorptions by all the major bands of H_2O, as well as those by several minor bands, are shown in Figure 7.1 where the bands of CO_2 and O_3 also appear.

TABLE 7.1   VIBRATION MODES AND TRANSITION PARAMETERS OF THE H_2O MOLECULE

Mode	Symbol	v (Hz)	λ (μm)	$\tilde{v}$ (cm^{-1})
Symmetric stretch[a]	v_1	1.097×10^{14}	2.734	3657.05
Bending	v_2	4.784×10^{13}	6.270	1594.78
Antisymmetric stretch[a]	v_3	1.127×10^{14}	2.662	3755.92
Harmonic of v_2	$2v_2$	9.464×10^{13}	3.170	3154.57
Combination of v_2 and v_3	$v_2 + v_3$	1.600×10^{14}	1.875	5332

[a]These bands overlap sufficiently that they are often treated as one band centered at 3700 cm^{-1} = 2.70 μm.

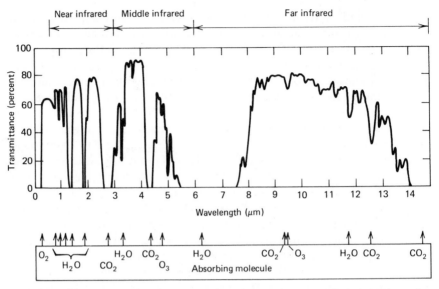

Figure 7.1. Transmittance of the atmosphere for a 6000-ft horizontal path at sea level containing 17 mm of precipitable water. From Hudson (1969) and Gebbie et al. (1951).

7.1.2. Computed Data

Readers should be aware of the numerous calculations of absorption line parameters that are available. Calculations have been made for the vibration–rotation lines of water vapor in the 2.7-μm region by Gates et al. (1964), and in the 1.9- and 6.3-μm bands by Benedict and Calfee (1967). Data for the pure rotational lines in the extreme infrared were calculated by Benedict and Kaplan; a small portion of their work is given by Goody (1964).

This earlier work, along with similar calculations for other gases, was continued at the Air Force Cambridge Research Laboratories (AFCRL), now the Air Force Geophysical Laboratory (AFGL). The result is a compilation of line parameters for all the major atmospheric gases, over the range 1 to 1000 μm (10,000 to 10 cm^{-1}), as reported by McClatchey et al. (1973). The gases considered are water vapor, carbon dioxide, ozone, nitrous oxide, carbon monoxide, methane, and oxygen. The absorption parameters are

Resonant frequency
Line intensity (strength)
Air-broadened half-width

Lower state energy
Vibrational quantum numbers
Rotational quantum numbers
Electronic state identifications
Isotope identifications

Data for more than 100,000 lines are presented, permitting high-resolution calculations of atmospheric transmittance and radiance. Because of the resolution characteristics, these compilations are known as HITRAN Computer Codes. Additions to the compilations, and information on codes for computations of optical paths, will be found in Clough et al. (1977), Rothman (1978a,b), and Rothman and McClatchey (1977). An updated version of HITRAN is described by Rothman (1981).

Similar compilations have been made by AFGL for the trace gases nitric oxide, sulfur dioxide, nitrogen dioxide, ammonia, nitric acid, chlorine monoxide, hydrogen halides, carbonyl sulfide, and formaldehyde. Information on these compilations is given by Rothman et al. (1978) and Rothman (1981). The line parameter data for both the major and the trace gases are recorded on magnetic tapes. Copies of the tapes are distributed by the National Climatic Center of NOAA, Digital Product Section, Federal Building, Asheville, NC 28801.

The detailed calculations of absorption and emission made possible by the HITRAN codes require considerable computer time. Also, the high resolution may not be warranted for applications where the spectral bandpass of the instrumentation is set by an optical filter. Of necessity, such filters are relatively wide-band. The need for only moderate resolution is met by the LOWTRAN Computer Codes, based on the initial work by McClatchey et al. (1972). These codes cover the nominal range 0.25–28.5 μm and include the extinction due to scattering by molecules and aerosol particles. Allowances are made for the effects of refraction and earth curvature on optical paths. The data are given in steps of 5 cm^{-1}, averaged over 20-cm^{-1} intervals. Progressive development of the LOWTRAN codes is described by Selby and McClatchey (1972, 1975) and Selby et al. (1976, 1978). Analyses of the functions and some relevant mathematical expressions are given by Pierluissi et al. (1979) and Gruenzel (1978). Comparisons of LOWTRAN predictions and measurements will be found in Haught and Cordray (1978, 1979), McClatchey and Kneizys (1979), and Ben-Shalom et al. (1981). Comments on the scattering functions included in the codes have been made by Cornette (1980) and Shettle et al. (1980). The LOWTRAN Computer Codes in card form are distributed by the NOAA activity identified previously.

Turning now to tabulated values of computed transmittance, Wyatt et al. (1964) provide such tabulations for water vapor over the wavenumber range

9950 to 1050 cm^{-1} (1.005 to 9.542 μm). The values are averaged over 100-cm^{-1} intervals, in steps of 50 cm^{-1}, for 1 atm pressure and 300 K, and for precipitable water content of 0.001–50 pw cm. These extensive tabulations are reproduced in Jamieson et al. (1963) and Valley (1965). Gibson and Pierliussi (1971) have derived an alternative set of transmittance parameters for the same range of wavelengths.

Tabulations similar to Wyatt et al. (1964), but covering the wider spectral range 0.3–13.9 μm, are given by Hudson (1969), who extended the work of Passman and Larmore (1956). He also gives altitude correction factors for reducing horizontal paths at high altitudes to their sea-level equivalents. It should be noted that the spectral range of these transmittance tables includes the large atmospheric window at 8–13 μm.

Bramson (1968) provides tables of transmittance for selected distances, and wavelengths from 1.0 to 5.3 μm and from 7.50 to 13.85 μm. He also tabulates attenuation coefficients for wavelengths in the range 2.0–14.0 μm, for 1.7 pw cm. Robinson (1966) presents graphs of fractional absorption for eight water vapor bands in the spectral range 0.72–6.3 μm, and for wide ranges of atmospheric pressure and precipitable water. Gates et al. (1963) have computed a set of transmission spectra for the water vapor band near 2.7 μm, in which they collect the vibrational modes v_1, v_3, and $2v_2$. The associated wavenumbers do not differ greatly, as may be seen from Table 7.1. Their computations are based on line parameter data generated by Gates et al. (1963) and are presented graphically. Eldridge (1967) gives values of water vapor transmission for the wavelength range 0.20–1.3 μm as a function of precipitable water.

The development of long-wave lasers and detectors is making the extreme-infrared and millimeter regions usable. Bastin (1966) has computed the absorption due to water vapor over the spectral range 250 μm– 4 mm and presents the results graphically. Augason et al. (1975) have generated an absorption spectrum for the wavenumber interval 45–185 cm^{-1} and compared it to a measured spectrum obtained with an airborne Michelson interferometer. Kyle (1975) presents absorption and transmittance data in both tabular and graphical form for the board range 1–2600 cm^{-1} (1 cm–3.85 μm). The gases considered are water vapor, carbon dioxide, ozone, nitrous oxide, methane, and oxygen. The tabulations cover number of lines, line strengths, and integrated absorptances for specified wavenumber intervals. Graphs of transmittance, averaged over intervals of 20, 5, and 0.1 cm^{-1} are given. All of the data pertain to vertical columns from the top of the atmosphere down to altitudes of 54, 45, 40, 30, 14, and 4 km. Traub and Steir (1976) have made similar but less extensive calculations of transmittance over the wavelength range 2–1000 μm. The values, presented graphically, refer to columns from the top of the atmosphere down to altitudes of 41, 28, 14, and 4.2 km, and to an *optical air mass* (secant of zenith angle) of 2.0. Tables of emissivities and emitted fluxes for these slanted columns also are included.

7.1.3. Measurement Data

So many measurements of absorption by water vapor have been made, by numerous workers using diverse methods, that there is a general abundance of data. Nevertheless, searching is often required for information that will be of direct help in solving a specific problem. In this section, which can be only a summary, we single out those measurements that seem to have wide applicability and that exemplify the principles discussed earlier. A helpful review of absorption measurements is given by Hudson (1969) from the viewpoint of systems engineering. Goody (1964), Jamieson et al. (1963), and Zuev (1974) discuss many sets of measurements in considerable technical depth.

The work of Elder and Strong (1953) in the earlier years of infrared technology is a good example of applying analytical methods to atmospheric measurements. During that era the infrared region out to 15 μm customarily was divided into eight window regions. These are shown in Figure 7.2, where the centers of the absorption bands form the divisions between the windows. Elder and Strong analyzed the results of seven quite

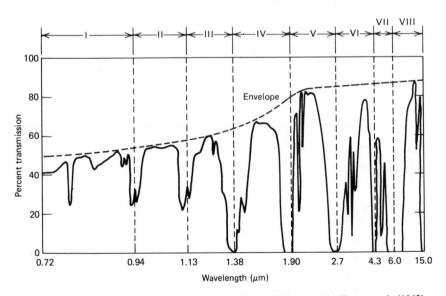

Figure 7.2. Infrared transmission windows in the atmosphere. After Kruse et al. (1963).

TABLE 7.2 TRANSMISSION WINDOW LIMITS AND EMPIRICAL CONSTANTS
FOR USE WITH EQ. (7.1)

Window	Wavelength Limits (μm)	c	t_0	τ equals 100% if pw is less than (mm)
I	0.70–0.92	15.1	106.3	0.26
II	0.92–1.1	16.5	106.3	0.24
III	1.1–1.4	17.1	96.3	0.058
IV	1.4–1.9	13.1	81.0	0.036
V	1.9–2.7	13.1	72.5	0.008
VI	2.7–4.3	12.5	72.3	0.006
VII	4.3–5.9	21.1	51.2	0.005
VIII	5.9–14		(Not treated)	

Source: From Hudson (1969).

different measurement programs and found that the effective transmittance
τ across a window is approximated by

$$\tau = -c \, \log(\text{pw mm}) + t_0 \qquad (7.1)$$

where pw mm is precipitable water in millimeters, and c and t_0 are constants
whose values are listed in Table 7.2. Values of transmittance calculated
from Eq. (7.1) are plotted in Figure 7.3. Langer (1957) extended the work
of Elder and Strong to take account of differing absorption characteristics
for small and large amounts of precipitable water. This extension and
associated data are described by Kruse et al. (1963).

Howard et al. (1956b) made many measurements of water vapor absorption, using a long-path test chamber in which the total pressure, water
vapor content, and temperature were varied. Similar measurements of
carbon dioxide absorption also were made, as summarized in Section 7.2.3.
For water vapor they found that the absorption data could be expressed by

$$\int A(\tilde{\nu})d\tilde{\nu} = cw^{1/2}(P + e_v)^k \qquad \text{(Weak band)} \qquad (7.2a)$$

$$\int A(\tilde{\nu})d\tilde{\nu} = C + D \, \log w + K \, \log(P + e_v) \qquad \text{(Strong band)} \quad (7.2b)$$

where the total pressure P and vapor pressure e_v are in mm Hg, and the
water vapor content w is in pw cm. The constants c and k in Eq. (7.2a) and
C, D, and K in Eq. (7.2b) have different values for each band, as given in
Table 7.3. The wavenumber that marks the transition from the weak regime

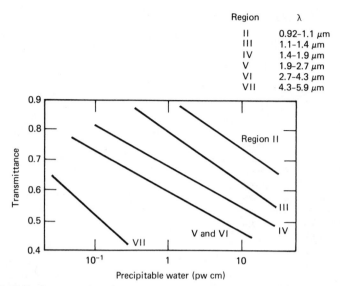

Region	λ
II	0.92–1.1 μm
III	1.1–1.4 μm
IV	1.4–1.9 μm
V	1.9–2.7 μm
VI	2.7–4.3 μm
VII	4.3–5.9 μm

Figure 7.3. Effective transmittance of infrared windows as a function of precipitable water. Data from Elder and Strong (1953).

(7.2a) to the strong regime (7.2b), for each band, is listed in the fifth column of the table. The entries in this column are the "strengths" of the absorption by the band, in energy units cm^{-1}.

Regarding the dual relationship shown by Eqs. (7.2), the authors state: "For some bands such as the 3.2-μm band, the weak-band relation (7.2a) was satisfactory for all data. . . . For most of the bands, however, low values of total absorption followed (7.2a) and large values of total absorption were given by the strong-band relation (7.2b); for such bands a transition value of total absorption could be selected to separate the range of total absorption best expressed by (7.2a) from that best expressed by (7.2b). In the immediate vicinity of the transition, an interpolation function must be used."

Assuming that all the absorption occurs within the band limits given in Table 7.2, an average fractional absorptance $\overline{A}$ for each band can be defined,

$$\overline{A} = \frac{1}{\bar{\nu}_2 - \bar{\nu}_1} \int A(\bar{\nu}) \, d\bar{\nu} \tag{7.3}$$

Taken together, Eqs. (7.2) and (7.3) have wide utility. If the spectral distribution of the flux incident on a given atmospheric path is known, these equations enable the computation of the fraction of total flux that is ab-

TABLE 7.3 EMPIRICAL RELATIONS FOR H_2O BANDS[a]

I. "Weak-Band" Fit: $\int A\nu d\nu = cw^{1/2}(P + p)^k$

H_2O Band (μm)	Band Limits (cm^{-1})	c	k	Transition $\int A_\nu d\nu$ (cm^{-1})
6.3	1150– 2050	356	0.30	160
3.2	2800– 3340[b]	40.2	0.30	500
2.7	3340– 4400	316	0.32	200
1.87	4800– 5900	152	0.30	275
1.38	6500– 8000	163	0.30	350
1.1	8300– 9300	31	0.26	200
0.94	10100–11500	38	0.27	200
3.7 (HDO)	2670– 2770	0.325	0.37	

II. "Strong-Band" Fit: $\int A_\nu d\nu = C + D \log w + K \log(P + p)$

H_2O Band (μm)	C	D	K
6.3	302	218	157
2.7	337	246	150
1.87	127	232	144
1.38	202	460	198

Source: Howard et al. (1956b).
[a]w in pw cm; P, p in mm Hg; logarithms to base 10. Band limits apply also to "strong" fit. "Weak" fit applies for $\int A_\nu \, d\nu$ less than transition value; "strong" fit for larger values.
[b]This band limit was arbitrarily chosen to "separate" overlapping bands.

sorbed within the appropriate spectral band, for the amount of water vapor selected. The fractional absorptance $\overline{A}$ given by Eq. (7.3) is equivalent to the radiometric property *absorptance* α defined in Appendix D. We then can write

$$\tau = 1 - \alpha = 1 - \overline{A} \qquad (7.4)$$

for the transmittance of the path. In principal summary of the foregoing, when the amount of absorber is varied over a very wide range, a two-part expression, Eqs. (7.2a) and (7.2b), gives a better fit to the absorption data than does Eq. (7.1).

The work by Howard et al. (1956b) described above was extended by Burch et al. (1963), who employed newer techniques and a method of accounting for the self-broadening effects of water vapor on the absorption lines. Such effects are noted in Section 6.1.4. Instead of total pressure P they dealt with an equivalent pressure P_e defined by

$$P_e = P + 4e_\nu$$

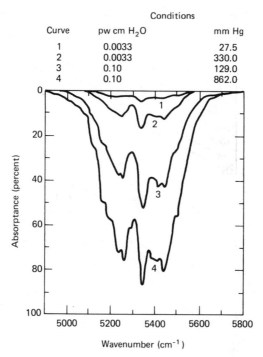

Figure 7.4. Spectral absorptance of H_2O band at 5332 cm^{-1} as a function of equivalent pressure and amount of absorber. From Burch et al. (1963).

to take account of self-broadening. They measured the spectral absorptance of the water vapor combination bands $\nu_2 + \nu_3$ at 5332 cm^{-1}, the ν_3 and ν_1 fundamentals near 3700 cm^{-1}, and the ν_2 fundamental at 1595 cm^{-1}.

One set of their spectral absorptance data is shown in Figure 7.4. The effects of increased water vapor and self-broadening are apparent in the upper three traces. Outstanding in the lower two traces is the effect of self-broadening alone. Here pw cm is held constant, and from the values of P_e it is seen that only P is varied. The entire set of traces shows strikingly the results of increased absorption. The far "wings" of the band are characterized by large values of the rotational quantum number J in the P and R branches, hence lesser populations according to Figure 4.8 and associated discussion. Thus the absorption is notably less at these rotational levels because of the smaller number of participating molecules. By contrast, near the band center the values of J are smaller and the rotational levels are more densely populated. Hence the center region of the band tends to be "absorbed out" and simple increases of the absorber amount produce smaller effects. The integrated or total absorptance for each of the three bands studied is shown in Figure 7.5 for various values of P_e and pw cm. The

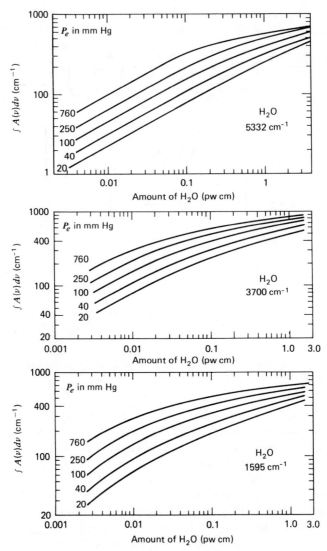

Figure 7.5. Total absorptances of the H_2O bands 5332, 3700, and 1595 cm^{-1} as functions of equivalent pressure and amount of absorber. From Burch et al. (1963).

smaller effects appear as decreasing slopes of the curves at higher values of the total absorptance. Although the data cited in the foregoing discussions are not recent, they are reliable and provide guidance in a field marked by complex relationships.

High-altitude investigations of the water vapor bands at 1.4, 1.9, and 6.3 μm have been made by Murcray et al. (1961) with balloon-borne spectrometers. At the ground Koepke and Quenzel (1978) have observed the spectral transmission in the very near infrared, using the sun as a source. Babrov and Casden (1968), working in the laboratory, measured the strengths of 12 lines in the ν_1 band at 2.73 μm (3663 cm^{-1}) and 30 lines in the ν_3 band at 2.65 μm (3774 cm^{-1}). The use of lasers in atmospheric research has brought about numerous measurements of absorption, particularly in the 8–13-μm window where a continuum of residual absorption exists. Measurements to investigate the continuum are reported by Bignell et al. (1963), McCoy et al. (1969), Nordstrom et al. (1978), Peterson et al. (1979), Shumate et al. (1976), and Tomasi et al. (1974). Roberts et al. (1976) have studied the problems of modifying the LOWTRAN code to provide a better fit to the measured absorption to the 8–13-μm window. Investigations of continuum absorption in the narrower window at 3.0–4.5 μm are described by Watkins and White (1977), Watkins et al. (1979), and White et al. (1978).

Measurements in the far- and extreme-infrared regions are noted. Detailed data on absorption in the 20–31.7-μm region are reported by Palmer (1957a), and in the 29–40-μm region by Palmer (1957b). Transmission functions for these regions have been developed by Palmer (1960). A supplement to this work is described by Stauffer and Walsh (1966) for the 14–20-μm region. Burch et al. (1968) investigated the long-wave expanse from 278 μm to 2 cm (from 0.5 to 36 cm^{-1}) both experimentally and theoretically. A study of rotational lines—positions, strengths, and widths—over the wavelength range 14.5 to 21 μm (692 to 475 cm^{-1}) was made by Izatt et al. (1969). Other investigations at high altitudes in the range 250–4000 μm (4 mm) are reported by Bastin (1966).

7.2. CARBON DIOXIDE ABSORPTION

Absorption by carbon dioxide, and studies of the possible climatic effects of the known increase in the concentration of this gas in the atmosphere, are receiving much present attention. As with the preceding section on water vapor, we now review the principal molecular factors, introduce the reader to several useful computations that have been made, and summarize some of the measurement data. In conclusion, attention is called to several additional investigations of CO_2 absorption.

7.2.1. Molecular Factors

The CO_2 molecule has a linear symmetric configuration, as shown in Figure 4.3, and in Table 4.2 we see that the rotational constant $B = 0.3902$. Because of its nuclear symmetry, the molecule has no permanent electric

TABLE 7.4   VIBRATION MODES AND TRANSITION PARAMETERS OF THE CO_2
MOLECULE

Mode	Symbol	ν (Hz)	λ (μm)	$\bar{\nu}$ (cm^{-1})
Symmetric stretch[a]	ν_1	—	—	—
Bending	ν_2	2.00×10^{13}	15.0	667
Antisymmetric stretch	ν_3	7.05×10^{13}	4.26	2349
Combination band[a]	$\nu_1 + \nu_3 - 2\nu_2$	7.28×10^{13}	4.12	2429

[a]Mode ν_1 normally inactive. Combination mode occurs when the ν_2 transition originates at
level $v = 2$ or higher.

dipole moment, hence no pure rotation spectrum. In two of its three vibrational modes, however, it acquires an oscillating dipole moment, which
is effective for rotational as well as vibrational transitions. This creates a
strong vibration–rotation band at each of the two fundamental frequencies
and also weaker bands at the overtone and combination frequencies.

The vibrational modes and parameters of the CO_2 molecule are listed in
Table 7.4, derived from Table 4.3 and Figure 4.14. The symmetric stretch
mode ν_1 is radiatively inactive at its fundamental because of nuclear and
vibrational symmetries. Its frequency is about twice that of the ν_2 fundamental, and it does interact with the other modes at higher vibrational
levels, which creates combination frequencies and resulting bands. The
bending mode ν_2 is degenerate and really consists of ν_{2a} and ν_{2b} vibrations
at the same frequency, as discussed near the end of Section 4.3.4. This is
l-type doubling and affects the quantum selection rule for rotation so that
$\Delta J = 0, \pm 1$. The ν_2 fundamental absorptive transition is represented by
$01^00 \leftarrow 00^00 \leftarrow$ and is a perpendicular vibration; therefore, the band has a
Q-branch corresponding to $\Delta J = 0$, as illustrated in Figure 5.3. The ν_3
fundamental is represented by $00^01 \leftarrow 00^00$ and is a parallel vibration. Thus
the rotational selection rule is simply $\Delta J = \pm 1$, and the Q-branch does not
appear. Several isotopes of CO_2 are present in the atmosphere; each one
has different fundamental frequencies as listed in Table 7.5.

Absorptions near 15 μm by the major bands of CO_2 have been investigated thoroughly because of their climatic importance. The band centers and several molecular factors for these bands are given in Table 7.6.
The first three entries represent the fundamental ν_2 vibration for each of the
isotopes listed in Table 7.5. Each absorptive transition takes place from the
initial level $v'' = 0$ to the next upward level $v' = 1$. This initial level has by

TABLE 7.5   FUNDAMENTAL VIBRATION FREQUENCIES OF ATMOSPHERIC CO_2 ISOTOPES

Species	Percentage Abundance	Band Center (cm^{-1})	
		ν_2	ν_3
$^{12}C^{16}O^{16}O$	98.420	667.40	2349.16
$^{13}C^{16}O^{16}O$	1.108	648.52	2283.48
$^{12}C^{16}O^{18}O$	0.408	662.39	2333

Source: Goody (1964).

far the greatest molecular population at terrestrial temperatures, as can be appreciated from Eq. (4.42). A study of the second column discloses that the remaining entries refer to transitions originating at levels higher than $v'' = 0$. In many cases they are multimode, involving transitions among several modes, including the normally inactive ν_1 vibration. Such simultaneous transitions are largely the result of the anharmonicities discussed in

TABLE 7.6   BAND STRENGTHS AND TRANSITION PARAMETERS FOR THE CO_2 BANDS NEAR 15 μm AT 300 K

Isotopic Species	Transition[a]	Band Center (cm^{-1})	(μm)	Molecular Band Strength (cm)	$\mathscr{E}''$ (cm^{-1})
$^{12}C^{16}O^{16}O$	00^00-01^10	667.40	15.0	7.89×10^{-18}	0.0
$^{13}C^{16}O^{16}O$	00^00-01^10	648.52	15.4	7.9×10^{-20}	0.0
$^{12}C^{18}O^{16}O$	00^00-01^10	662.39	15.1	3.7×10^{-20}	0.0
$^{12}C^{16}O^{16}O$	01^10-02^00	618.03	16.2	1.75×10^{-19}	667.4
—	01^10-10^00	720.83	13.9	2.3×10^{-19}	667.4
—	01^10-02^20	667.76	15.0	6.2×10^{-19}	667.4
—	02^00-03^10	647.02	15.5	4.2×10^{-20}	1285.43
—	02^00-11^10	791.48	12.6	8.2×10^{-22}	1285.43
—	02^20-03^10	597.29	16.7	5.83×10^{-21}	1335.16
—	02^20-11^10	741.75	13.5	5.2×10^{-21}	1335.16
—	02^20-03^30	668.3	15.0	3.2×10^{-20}	1335.16
—	10^00-03^10	544.26	18.4	1.64×10^{-22}	1388.19
—	03^30-04^20	581.2	17.2	1.56×10^{-22}	2003.28
—	03^30-12^20	756.75	13.2	2.2×10^{-22}	2003.28
—	03^10-12^20	828.18	12.1	1.8×10^{-23}	1932.45
—	03^10-12^00	740.5	13.5	5.2×10^{-22}	1932.45
—	04^40-13^30	769.5	13.0	1.5×10^{-23}	2674.76

[a]In this listing the term symbol for the lower state appears first.
Source: Goody (1964).

Section 4.3.4. In the fifth column the constant-valued zero-point energy is not included, since only energy differences are important in transitions.

The remaining major bands of CO_2 are near 4.3 μm, and these are sufficiently strong to render the atmosphere opaque in this region. The ν_3 and combination modes and frequencies of the $^{12}C^{16}O^{16}O$ isotope are given by the last two entries in Table 7.4. These overlap the ν_3 mode of the $^{13}C^{16}O^{16}O$ isotope at 2283 cm^{-1}, whose percentage abundance is given in Table 7.5. The combined effects of the three bands are often considered in terms of one aggregate band. Each of the three is a parallel band and thus has no Q-branch. Other weaker bands appear between 1.4 and 15 μm; some of these are covered by the discussion in Section 7.2.3.

7.2.2. Computed Data

Since carbon dioxide has no pure rotation spectrum, and its electronic bands appear to be of minor importance to atmospheric optics, our interest is confined to the infrared region out to the vibration–rotation bands near 15 μm. The HITRAN and LOWTRAN compilations and computer codes of absorption parameters, discussed in Section 7.1.2, apply throughout this region. Stull et al. (1964) provide tabulated values of computed transmission by CO_2 in this region; these tables are the counterpart of the H_2O transmission tables by Wyatt et al. (1964). The CO_2 tables cover the wavenumber range 9950 to 550 cm^{-1} (1.005 to 18.2 μm). The transmittance values are averaged over intervals of 50 cm^{-1} for wide ranges of path length expressed in terms of the basic quantity atm cm of CO_2. These tables are reproduced in Jamieson et al. (1963) and Valley (1965). Similar tabulations are given by Hudson (1969) for the spectral range 7.0–13.9 μm, applicable to a horizontal path at sea level. Included are altitude correction factors for reduction of an actual horizontal path to an equivalent sea-level path. Robinson (1966) presents graphs of fractional absorption for the bands centered at 1.6, 2.0, 2.7, and 4.3 μm, for a range of about 0.1–10,000 atm cm of CO_2.

Recent years have seen major accomplishments in the "inverting" of satellite-borne radiometric measurements of the atmosphere to derive profiles of temperature and gas concentration. To this purpose Drayson (1965, 1966) had an early look at transmission in the CO_2 bands near 15 μm. Drayson et al. (1967, 1968) provide tabulations of line and band strengths in this region. Also in this connection, Roney et al. (1978) examine the accuracy of the absorption line compilations by McClatchey et al. (1973). Wang (1978) analyzes the effect of temperature on transmission in the 15-μm region.

7.2.3. Measurement Data

The absorption by carbon dioxide in eight separate bands was measured by Howard et al. (1956a) in their investigations of infrared transmission. The water vapor portion of this program is summarized in Section 7.1.3. For carbon dioxide they found that most of the absorption data could be expressed by

$$\int A(\bar{\nu})d\bar{\nu} = cw^{1/2}(P + p)^k \qquad \text{(Weak band)} \qquad (7.5a)$$

$$\int A(\bar{\nu})d\bar{\nu} = C + D \log w + K \log(P + p) \qquad \text{(Strong band)} \quad (7.5b)$$

where the total pressure P and partial pressure p of CO_2 are in mm Hg and the amount w of CO_2 is in atm cm. The constants c and k in Eq. (7.5a) and C, D, and K in Eq. (7.5b) have different values for each band, as shown in Table 7.7. The absorption strength, in energy units of cm^{-1}, that marks the transition from the weak regime (7.5a) to the strong regime (7.5b), for each band, is listed in the fifth column of the table. Just as for the case of water vapor, an average fractional absorptance $\bar{A}$ for each band is defined by

$$\bar{A} = \frac{1}{\bar{\nu}_2 - \bar{\nu}_1} \int A(\bar{\nu}) \, d\bar{\nu} \qquad (7.6)$$

analogously to Eq. (7.3). The transmittance then is

$$\tau = 1 - \bar{A}$$

as with Eq. (7.4).

Burch et al. (1962b) made further measurements of absorption for all of the carbon dioxide bands. As in their work discussed in Section 7.1.3, they employed as the pressure parameter an equivalent pressure P_e, defined for the present case by

$$P_e = P + 0.30p$$

where P is total pressure and p is the partial pressure of CO_2, both expressed in mm Hg. In extending the work of Howard et al. (1956a) they obtained many curves of CO_2 spectral absorptance similar to those shown by Figure 7.4 for H_2O. The behavior of CO_2 absorptance was very similar to that of H_2O in that the absorptance at first increases steadily with

TABLE 7.7   SUMMARY OF EMPIRICAL RELATIONS[a]

I. "Weak-Band" Fit: $\int A_v \, dv = cw^{1/2}(P + p)^k$

CO^2 Band (μm)	Band Limits (cm^{-1})	c	k	Transition $\int A_v dv$ (cm^{-1})
15	550– 800	3.16	0.44	50
5.2	1870–1980[b]	0.024	0.40	30
4.8	1980–2160[b]	0.12	0.37	60
4.3	2160–2500	—	—	50
2.7	3480–3800	3.15	0.43	50
2.0	4750–5200	0.492	0.39	80
1.6	6000–6550	0.063	0.38	80
1.4	6650–7250	0.058	0.41	80

II. "Strong-Band" Fit: $\int A_v dv = C + D \log w + K \log(P + p)$

CO_2 Band (μm)	C	D	K
15	− 68	55	47
4.3	27.5	34	31.5
2.7	−137	77	68
2.0	−536	138	114

Source: Howard et al. (1956a).
[a] w in atm cm; P, p, in mm Hg; logarithms are to base 10. Band limits apply also to "strong" fit. "Weak" fit applies for $\int A_v dv$ less than "Transition $\int A_v dv$"; "strong" fit for higher values.
[b] These band limits were arbitrarily chosen to "separate" overlapping bands.

increases of absorber amount and pressure. This continues until the absorptance near band center is nearly complete, when further increases of these parameters produce further absorptions only toward the wings of the bands. This overall decreasing effect, corresponding to the changeover from Eq. (7.5a) to Eq. (7.5b) as the amount of CO_2 is increased, is clearly shown for the 4.3-μm band by Figure 7.6.

Further work is noted. Absorption in the band between 8000 and 10,000 cm^{-1} (1.25 and 1 μm) was investigated by Burch et al. (1968), and in the band between 1100 and 1835 cm^{-1} (9.1 and 5.5 μm) by Burch and Gryvnak (1971). Measurements of integrated band strength were made by Ellis and Schurin (1969) for the 4.82- and 5.17-μm regions, and by Schurin and Ellis (1968) for the 1.43-, 1.6-, and 2.0-μm regions. Aronson et al. (1975) employed a tunable diode laser to study the rotational lines of the Q-branch of the v_2 band at 15 μm. Eng et al. (1977) likewise determined the center of the v_3 band of the isotope $^{14}C^{16}O^{16}O$ near 4.5 μm. Monchalin et al. (1977)

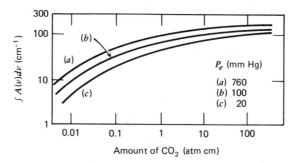

Figure 7.6. Total absorptance of the CO_2 band at 2350 cm^{-1} as a function of equivalent pressure and amount of absorber. From Burch et al. (1962b).

employed an interferometer having a resolving power of nearly 10^9 to measure the wavelength of the $R(14)$ line in the 9.4-μm combination band. By using the known frequency of this line, they then determined a value for the speed of light in close agreement with values obtained from independent measurements. This work is an interesting update of that by Plyler et al. (1955) who employed the molecular constants of the CO molecule in their determination of the speed of light, as noted in Section 4.2.3.

7.3. OXYGEN AND OZONE ABSORPTION

Considerable present interest is directed to the ultraviolet absorptions by high-altitude oxygen and ozone and their consequent environmental roles. The O_2 molecule is a symmetrical diatomic species having but one mode of vibration, as in Figure 4.14. Because of the symmetry it has neither a permanent nor an oscillating electric dipole moment, hence is radiatively inactive with respect to simple rotation and vibration. However, some radiative activity does occur in the infrared, visible, and ultraviolet (importantly), due to transitions between electronic states. Several of these states are shown in Figure 4.19. Absorptions in the middle ultraviolet between 175 and 195 nm have been investigated by Hudson and Carter (1968), and near 122 nm by Gailey (1969). Magnetic dipole transitions among the three lowest states lead to weak absorption bands at 1.27, 1.06, and 0.762 μm. Goody (1964), Wark and Mercer (1965), and Burch and Gryvnak (1969) should be consulted in this regard.

The O_3 molecule is an asymmetric top, as in Figure 4.3, and its three rotational constants are listed in Table 4.2. The vibrational modes and parameters are given in Table 7.8, from Figure 4.14 and Table 4.3. These produce vibration–rotation bands of moderate strength at 4.75 and 14.2 μm

TABLE 7.8   VIBRATION MODES AND TRANSITION PARAMETERS OF THE O_3 MOLECULE

Mode	Symbol	ν (Hz)	λ (μm)	$\bar{\nu}$ (cm^{-1})
Symmetric stretch	ν_1	3.33×10^{13}	9.01	1110
Bending	ν_2	2.12×10^{13}	14.2	705
Antisymmetric stretch	ν_3	3.13×10^{13}	9.59	1043

and a very strong band at 9.6 μm. The electronic absorption bands in the ultraviolet, known as the *Hartley bands,* are centered at about 255 nm, as in Figure 7.7. Here the O_3 molecule has a maximum cross section of about 1.1×10^{-17} cm^2. These bands are responsible for shielding the earth's surface from harmful solar ultraviolet. Effectiveness of the shielding may be realized from the fact that a total ozone amount of 0.35 atm cm represents approximately 9.4×10^{18} molecules in a vertical column having a cross section of 1 cm^2. The absorption coefficient of the column is then about 103, and the resulting transmittance of the column is near 10^{-45}. Other absorptions occur in the weak Huggins bands at 310 nm, and in the weak Chappuis bands extending from 450 to 750 nm. In addition to its strong absorption of ultraviolet, the infrared absorption at 9.6 μm by the ozone layer is seen to be an agent in the temperature rise of the stratosphere.

Measurements of the ultraviolet absorption by ozone have been made by Inn and Tanaka (1953), Strong (1941), Tanaka et al. (1953), and Vigroux (1953). In the infrared Goldman et al. (1970b) have measured the absorption in the interval 9–10 μm, over very long paths observed from a balloon. In an associated effort Goldman (1970) computed the band parameters for this interval. El-Sherbiny et al. (1979) employed a tunable diode laser and high-resolution spectroscopy to investigate the ν_3 band at 9.6 μm. Young and Bunner (1974) studied the absorption of CO_2 laser lines at wavelengths near 9.4 μm. Other measurements of the 9.6-μm band have been made by Kaplan et al. (1956) and Walshaw (1957).

7.4.   ABSORPTION BY MINOR GASES

The minor gases whose absorptions are considered here are methane, carbon monoxide, nitrogen dioxide, nitric oxide, nitrous oxide, and nitric acid vapor. Except in regions of pollution these gases usually are present in only the trace amounts indicated in Tables 1.1 and 1.2. Information beyond

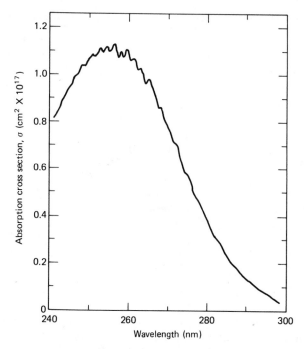

Figure 7.7. Absorption cross-section of O_3 molecule in the Hartley bands centered at 255.3 nm. The bands, indicated by the fluctuations on the curve, overlap and are superposed on a strong continuum. From Goody (1964).

that given here can be found in Goldberg (1954), Goody (1964), and Zuev (1974). Brinkman et al. (1966) treat the absorptions of these gases in considerable detail, from the viewpoint of upper-atmosphere physics, and provide abundant references. Many absorption lines of these gases are shown and identified in the spectral atlas by Goldman et al. (1978).

7.4.1. Methane

The CH_4 molecule has a spherical top configuration, as in Figure 4.2, and three equal rotational constants designated by the single value under B in Table 4.2. It has no pure rotation spectrum, but does have four fundamental vibration modes. Of these, only ν_3 and ν_4 are active in the infrared, as listed in Table 4.3, and create vibration–rotation bands. Overtone and weak combination bands also exist, the latter involving interactions with the normally inactive ν_1 and ν_2 vibrations. Most of these bands are listed in Table 7.9, where the symbolism for transitions allows for the four vibrational modes.

TABLE 7.9   METHANE BANDS OBSERVED IN THE SOLAR SPECTRUM

Band Center (cm^{-1}) (μm)		Transition[a]	Molecular Band Strength at STP (cm)
6005	1.67	0000–0020	3.6×10^{-20}
5861	1.71	0000–1101	—
5775	1.73	0000–0111	—
4420	2.20	0000–0110	—
4313	2.32	0000–0011	—
4216	2.37	0000–1001	—
4123	2.43	0000–0102	—
3019	3.31	0000–0010	1.26×10^{-17}
3823	3.55	0000–0101	—
2600	3.85	0000–0002	1.04×10^{-17}
1306	7.66	0000–0001	6.9×10^{-18}

[a]In this listing the term symbol for the lower state appears first.
Source: Goody (1964).

The ν_3 band at 3020 cm^{-1} (3.3 μm) has been studied by Allen and Plyler (1957), Finkman et al. (1967), and Seeley and Houghton (1961), and the $2\nu_3$ band near 6000 cm^{-1} by Margolis et al. (1974). Hoover et al. (1967) have investigated the absorption in the region where the ν_4 band of CH$_4$ at 1306 cm^{-1} is superposed on the ν_1 fundamental band of N$_2$O at 1285 cm^{-1}. Burch and Williams (1962b) measured the total absorptance of CH$_4$ in the ν_3 and ν_4 bands for wide ranges of absorber concentration w and equivalent pressure P_e. Some of their results are shown in Figure 7.8, where P_e is defined by

$$P_e = P + 0.30p \qquad (3020\text{-cm}^{-1} \text{ band})$$

$$P_e = P + 0.38p \qquad (1306\text{-cm}^{-1} \text{ band})$$

In the cited work, the authors derive analytical expressions for the total absorptance, which is the ordinate scale in the figure.

7.4.2.   Carbon Monoxide

The CO molecule has a simple diatomic configuration, as shown in Figure 4.14, and the single rotational constant listed in Table 4.2. Because of its permanent dipole moment, it has a pure rotational spectrum, and the transition $J'' = 0$ to $J' = 1$ has been observed in the microwave emission

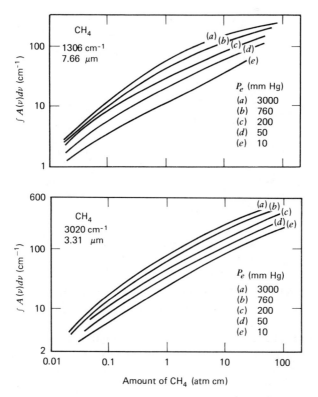

Figure 7.8. Total absorptances of CH_4 in the ν_3 band at 3020 cm^{-1} and the ν_4 band at 1306 cm^{-1} as functions of equivalent pressure and amount of absorber. From Burch and Williams (1962b).

spectrum of galactic CO, mentioned in Section 5.1. The CO molecule has a single fundamental mode ν_1 at 2143 cm^{-1} (4.67 μm), as listed in Table 4.3. Because of its structural simplicity, this molecule has been studied thoroughly and has served as an example of molecular parameters in previous chapters.

The ultraviolet band spectra of CO are treated extensively by Herzberg (1961) and Watanabe (1958). A set of vibrational band progressions for the $^1\Pi \leftrightarrow {}^1\Sigma^+$ electronic transition is described in Section 5.3, using the values listed in Table 5.1. The infrared band near 2143 cm^{-1} has been investigated by many workers. Strengths and widths of its rotational lines have been measured by Hoover and Williams (1969) and Shaw and France (1956). Shaw and Houghton (1964) calculated the total band absorptance as a function of pressure and amount of absorber and studied the effects of line width on total absorptance. Vanderwerf and Shaw (1965) measured the

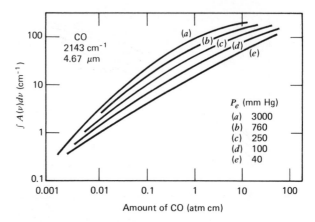

Figure 7.9. Total absorptance of CO in the ν_1 band at 2143 cm^{-1} as a function of equivalent pressure and amount of absorber. From Burch and Williams (1962b).

effect of temperature on total absorptance, and Goldman et al. (1973c) and Seeley and Houghton (1961) studied the band with an airborne spectrometer.

Burch and Williams (1962b) measured the total absorptance of the band for wide ranges of equivalent pressure P_e and amount of absorber. One set of their data is shown in Figure 7.9, where P_e is defined by

$$P_e = P + 0.02p$$

It is seen that the general features of the curves are similar to those in Figure 7.8 as well as to those in Figures 7.5 and 7.6. The authors discuss a number of empirical relations derived during the laboratory work.

7.4.3. Nitrogen Gases

The nitrogen gases present in the atmosphere are nitrogen dioxide, nitric oxide, nitrous oxide, and nitric acid vapor. Their trace amounts are indicated in Tables 1.1 and 1.2. All of these gases have spectra in the infrared and, in some cases, in the ultraviolet. Many of the bands are strong but, owing to the small presence of the gas, the absorptions have only a minor influence on atmospheric temperature and terrestrial heat balance. Interest centers on these constituents because of their roles in the subtle photochemistry of the atmosphere, for example, in affecting the equilibrium value of total ozone, or as active contributors to photochemical smog—a particularly irritating species of pollution. We now consider each of these gases.

The NO_2 molecule has a linear asymmetric structure, the rotational constant is listed in Table 4.2, and the fundamental vibration frequencies are given in Table 4.3. Each of these gives rise to a vibration–rotation band, the strongest of which is the ν_3 band centered at 1618 cm^{-1} (6.18 μm). This is near the center of the ν_2 band of H_2O at 1595 cm^{-1}. Additional bands associated with electronic transitions produce absorptions in the near ultraviolet and visible regions. Little work on these bands in the atmosphere has been reported. Takeuchi et al. (1978) have investigated the use of a DAS (differential absorption scattering) lidar for finding the spatial distribution of NO_2. O'Shea and Dodge (1974) measured the absorption coefficients at argon-ion laser wavelengths from 0.4965 to 0.5145 μm. Hsu et al. (1978) provide a spectral atlas for the interval 5530–6480 Å, listing 19,000 lines and carrying over 500 references. Shafer and Young (1976) have measured the total absorptance of the $2\nu_3$ band at 3.44 μm. Goldman et al. (1970a) detected the ν_3 band from 1550 to 1640 cm^{-1} with a balloon-borne spectrometer and provide a spectral trace of the absorption.

The NO molecule has a linear asymmetric structure, the rotational constant is in Table 4.2, and the fundamental frequency is in Table 4.3. Nitric oxide has extensive spectra in the ultraviolet but only minor activity in the infrared. Tajime et al. (1978) and Watanabe et al. (1967) have measured the absorptions and photoionizations at 2260 Å and in the interval 580–1350 Å. Mantz et al. (1976) report emission measurements between 5 and 7 μm, and Richton (1976) has measured the transmittance for CO laser lines near 1900 cm^{-1}.

The N_2O molecule has a linear asymmetric structure, the single rotational constant is in Table 4.2, and the three fundamental frequencies are listed in Table 4.3. There are several absorption bands and continua in the ultraviolet; absorption and photoionization coefficients have been measured by Cook et al. (1968). Numerous absorption bands formed by the fundamental, overtone, and combination frequencies exist in the infrared. Goody (1964) lists these bands and their molecular parameters. Tidwell et al. (1960) have investigated the rotational lines of several such bands to quantum numbers near $J = 50$. Abels and Shaw (1963) measured the dependences of total absorptance on pressure and amount of absorber for the ν_3 band near 4.5 μm, and Yale et al. (1968) and Burch and Williams (1962a) investigated the band strength. Seeley and Houghton (1961) and Goldman et al. (1970c, 1973a) have studied this band at various altitudes and analyzed its spectrum closely.

Nitric acid vapor was discovered in the atmosphere above 20 km by Murcray et al. (1969a), and since then it has received much attention. The HNO_3 molecule has an approximately planar configuration, with absorption bands at 5.9, 7.5, 11.3, and 22 μm. Laboratory measurements of its absorptance, and comparisons with atmospheric spectra, have been made by Goldman et al. (1971), Rhine et al. (1969), and Schmidt et al. (1974). Farrow et al. (1979) have derived parameters for some of its lines by measuring the transmittance of CO laser lines near 5.84 μm. Dozens of the HNO_3 absorption lines are identified in the spectral atlas prepared by Goldman et al. (1978). Additional studies of this intriguing constituent of the upper atmosphere are reported by Goldman et al. (1980, 1981) and Lado-Bardowsky and Amat (1979).

7.5. ATMOSPHERIC TRANSMISSION

The renowned set of transmission measurements by Gebbie et al. (1951) were the most comprehensive to that time, and they have retained much of that distinction. Their work covered the customarily used infrared region from 1 to 14 μm. The 6000-ft (1.8-km) optical path employed was about 100 ft above sea level, and for most of its distance the path lay over the sea along the coast of Scotland. Measurements were made under a variety of meteorological conditions, but the reduced data refer to a water vapor content of 17 pw mm along the path and a visual transmission of 60% per nautical mile (1.85 km) at 0.61 μm. The attenuation at this wavelength was due almost entirely to scattering by haze. A summary result of their work is shown by the transmission spectrum in Figure 7.1.

Taylor and Yates (1956) measured transmittance horizontal paths at Chesapeake Bay for lengths of 1000 ft (0.3 km), 3.4 mi (5.5 km), and 10.1 mi (16.2 km). The amounts of water vapor along the paths, respectively, were 1.1, 13.7, and 52 pw mm. Because of their wide applicability, these detailed transmittance curves, with the measurement resolution indicated thereon, are reproduced at a large scale in Appendix G. This scale permits a fairly accurate measurement of the area under a curve, by either a coordinate overlay or a planimeter, for any selected bandpass in a region of interest. Thus the average transmittance for any desired interval can be found easily. Yates and Taylor (1960) report later measurements of this type at Chesapeake Bay and Hawaii.

Other measurements are noted. Streete et al. (1967) measured the transmittance of a 25-km horizontal path along the Florida coast, for the spectral region 0.68–4.86 μm. The water vapor in the path varied between 9 and 39 pw cm. Matching solar spectra were obtained, and the two sets of traces are aligned for comparison. Similar measurements over the same path, but for the region 0.56–10.7 μm, are reported by Streete (1968), for values of water vapor ranging from 21.5 to 43.3 pw cm. The author compares his results to those of Taylor and Yates (1956), and discusses transmittance with respect to the nominal atmospheric windows shown in Figure 7.2. Haught and

Dowling (1977) have measured the transmittance at Chesapeake Bay over a 5-km path, for the wavenumber interval 2480–2800 cm^{-1}. They compare their results with values calculated from the line parameter compilation by McClatchey et al. (1973), supplemented with models from Burch et al. (1971). Very reassuring comparisons are shown.

Turning now to higher-altitude conditions, Gates and Harrop (1963) scanned the solar spectrum from a Denver site which was 1615 m (5290 ft) above sea level. Their observations covered the spectral region 1.0–12.5 μm and yielded absorption coefficients and identification of species. Measurements of the solar spectrum from balloon and mountain sites are reported by Kondratyev et al. (1964, 1965) and Murcray et al. (1960. Murcray et al. (1965) have measured solar spectra over the wavenumber interval 1590–2500 cm^{-1} and studied the variation of absorption with altitude. They derived values of integrated absorptance for three bands in this interval and compare their results with theoretical values. Murcray et al. (1969b) report similar but more extensive measurements in the 4–14.3-μm region and show many spectral traces having transmittance as the ordinate. Atmospheric absorption in the submillimeter region at wavenumbers less than 125 cm^{-1} (greater than 80 μm) have been made at high altitudes by Eddy et al. (1969) and Nolt et al. (1971).

7.6. ATMOSPHERIC INFRARED EMISSION

Until recently, the measurement of infrared emission seemed to receive less attention than the measurement of absorption, although the two processes are of equal significance. Developments of sensitive radiometers and spectrometers that sense radiant flux in all spectral regions, however, now enable emission data to be gathered that are of primary importance in understanding atmospheric structure and composition. With some simplification, atmospheric emission can be placed in two categories: (1) thermal or infrared emission, principally by the lower atmosphere, and (2) airglow and auroral emission, mainly in the visible, by the mesosphere and ionosphere.

Airglow, sometimes called nightglow because its faintness usually precludes daytime observation, is a photochemical luminescence of quasi-steady value over middle and low latitudes. The processes leading to airglow are not fully defined but, broadly said, airglow appears to be the radiative release of latent molecular energy stored during the daytime. Auroras are sporadic radiant emissions usually occurring over middle and high latitudes. Various theories relate auroras to an increased flux of charged solar particles and the usually accompanying magnetic storms.

Airglow and auroras are outside the scope of this book. Succinct accounts will be found in Craig (1965) and Fleagle and Businger (1963), and comprehensive treatments in Chamberlain (1958, 1961).

Here we restrict our view to infrared emission, particularly to measurement data that illustrate the processes. We call attention in passing, however, to the "retrieval techniques" being applied to the spectral radiance data now gathered in vast amounts by satellites as they scan the earth–atmosphere system. The objective of the techniques, known generally as *inversion of measurements,* is to retrieve or extract profiles of temperature and absorber (emitter) concentration from the surface upward. The objective, almost the problem, is stated by Rodgers (1976) in a tutorial treatment: "Given a measurement or series of measurements of thermal radiation emitted by an atmosphere, the intensity and spectral distribution of which depend on the state of the atmosphere in a known way, deduce the state of the atmosphere." Other introductory treatments of measurement inversion are given by Conrath (1972), Gille (1968), Kaplan (1959), Wark (1970), and Yamamoto (1961). Deepak (1977) brings a group of experts to bear on just about every aspect of measurement inversion, including applications to aerosol scattering.

For our purpose the phenomenology of atmospheric infrared emission is shown rather clearly by the spectra in Figure 7.10, from the work by Sloan et al. (1955). In part (a) we see the CO_2 absorption bands at 4.3 and 15 μm, the great H_2O band at 6.3 μm, and the O_3 band at 9.6 μm. The additional band at 2.7 μm is due to overlapping bands of CO_2 and H_2O. Minor bands and groups of lines can be discerned beyond 7.5 μm. Part (b) is the sky spectrum at the noon hour on a clear day of moderate humidity, with the ground-level air temperature at 20°C. The topmost curve is the spectrum of a blackbody at 21.5°C. The dotted curve represents a correction for CO_2 and H_2O absorptions within the spectrograph. Comparing now parts (a) and (b), it is seen that for every minimum in the solar spectrum caused by atmospheric absorption there is a corresponding maximum in the atmospheric emission spectrum. The coincidences are striking and are repeated in the other panels of the figure. Part (c) is the spectrum of the clear sky at night, and its similarity to (b) indicates that the effects of scattered sunlight are not present significantly in the daytime spectrum at the wavelengths employed. Finally, Figure 7.11 shows the increased emission, particularly in the window regions, as the zenith angle of observation is increased, thereby producing a greater optical thickness of the atmospheric path.

Very detailed emission spectra of the zenith sky have been obtained by Arnold and Simmons (1968), one example of which is given in Figure 7.12. Not unexpectedly, the greater emissions occur at higher values of temper-

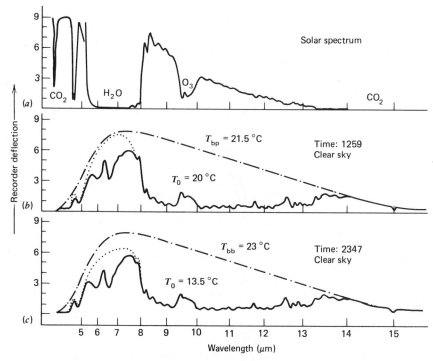

Figure 7.10. Emission spectra of the clear sky toward the zenith, with a solar spectrum for comparison. From Sloan et al. (1955).

ature and humidity. Outstanding are the emission continuum across the window region from 8 to 13 µm or so, and its increase with temperature and humidity. This continuum is, of course, the emission counterpart of the absorption continuum in this window mentioned near the end of Section 7.1.3.

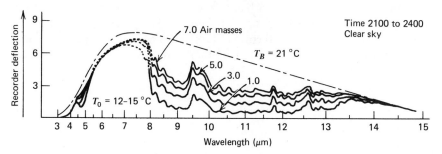

Figure 7.11. Emission spectra of the clear night sky observed at several zenith angles, shown in terms of air masses. From Sloan et al. (1955).

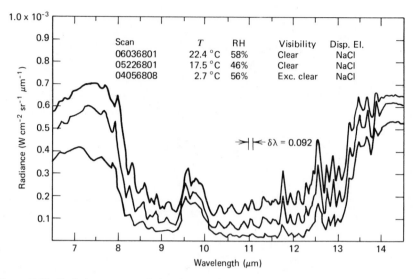

Figure 7.12. Emission spectra of the clear zenith sky observed for several conditions of temperature and relative humidity. From Arnold and Simmons (1968).

In conclusion we call attention to other work. An excellent discussion of infrared emission by the atmosphere is contained in Jamieson et al. (1963). Additional measurements of sky emission have been made by Sloan et al. (1956), beyond those discussed above under Sloan et al. (1955). Bell et al. (1960) report on the spectral radiance of the sky at wavelengths from 1 to 20 μm, and in particular they show its dependence on zenith angle. Bennett et al. (1960) investigated the distribution of total radiance over the sky in the region from 2 to 40 μm. Measurements from balloons have been made by Goldman et al. (1973b) in the 24–29μm region, by Jennings and Moorwood (1971) with an interferometer over the range 10–100 μm, and by Murcray et al. (1962) in various bands from 2 to 35 μm. Measurements from aircraft altitudes near 11 km are reported by Huppi et al. (1974) in the region between 700 and 2800 cm^{-1} (14.28 and 3.57 μm), and by Beckman and Harries (1975) and Carli et al. (1977) in the extreme infrared.

Appendix A

Important Physical Constants

Speed of light, vacuum	$c = 2.9979 \times 10^8$ m sec^{-1}
Charge of electron	$e = 1.6021 \times 10^{-19}$ C
Rest mass of electron	$m = 9.1091 \times 10^{-31}$ kg
Planck's constant	$h = 6.6256 \times 10^{-34}$ J sec
First radiation constant	$c_1 = 3.7415 \times 10^{-4}$ W cm^{-2}
Second radiation constant	$c_2 = 1.4388 \times 10^4$ μm K
Photon radiation constant	$c_1' = 1.8836 \times 10^{15}$ sec^{-1} cm^{-2}
Stefan–Boltzmann constant	$\sigma = 5.6697 \times 10^{-12}$ W cm^{-2} K^{-4}
Wien displacement constant	$\lambda T = 2.8978 \times 10^3$ μm K
Avogadro's number	$N_A = 6.0225 \times 10^{23}$ mol^{-1}
Molar volume of ideal gas	$V_A = 2.2414 \times 10^{-2}$ m^3 mol^{-1}
Loschmidt's number (STP)	$N_L = 2.6871 \times 10^{19}$ cm^{-3}
Boltzmann's constant	$\kappa = 1.3805 \times 10^{-23}$ J K^{-1}
Gas constant (universal)	$R. = 8.3143$ J mol^{-1} K^{-1}
Gas constant (dry air, STP)	$R = 2.8706 \times 10^2$ J kg^{-1} K^{-1}
Gas constant (water vapor)	$R_v = 4.6191 \times 10^2$ J kg^{-1} K^{-1}
Density of dry air (SPT)	$\rho = 1.2928$ kg m^{-3}
Mean molecular weight of air	$= 28.964$
Mean mass of air molecules	$m = 4.81 \times 10^{-26}$ kg molecule^{-1}
Acceleration of gravity (sea level)	$g = 9.8067$ m sec^{-2}
Standard atmosphere	$= 1.0133 \times 10^5$ N m^{-2}
(sea-level pressure)	$= 1.0133 \times 10^3$ mb
	$= 760$ mm Hg $= 29.213$ in. Hg

Source: Data from List (1966), USSA (1976), and Valley (1965).

Appendix **B**

Physical Units
and Conversions

The concepts and methods of mechanics form a unifying structure for the various divisions of physical science. Mechanics is basic and three quantities therein—*length, mass,* and *time*—were named by Gauss in 1832 as *absolute*. These appear to be elementary concepts which can be neither derived from one another nor resolved into anything more primitive. We may have an almost intuitive recognition of the fundamental nature of length, mass, and time, which we employ in perceiving the physical world.

As the science of physics progressed, several additional quantities were derived from the three fundamentals, and electrical quantities were added to form systems of units. In long ago 1861 the British Association for the Advancement of Science recommended the adoption of a centimeter–gram–second (CGS) system for specifying physical and electrical quantities. Three versions of the CGS system came into use, namely, the electrostatic (esu) and electromagnetic (emu) systems, and the Gaussian system consisting of mixed esu and emu quantities. The Giorgi or rationalized meter–kilogram–second (MKS) system was proposed in 1901 and slowly made headway. The unit *ampere* then was added for the electrical quantity, thus forming the MKSA system. These four systems and several intermediaries were in concurrent and overlapping use for many years. The resulting plethora of units has been deplored by many persons concerned with their use.

In 1954 the General Conference on Weights and Measures extended the rationalized MKSA system and established the International System of Units (Le Système International d'Unités), known as the SI system. SI is based on seven well-defined *base units,* which by convention are regarded as dimensionally independent. These form the first group in Table B.1. To these are added the supplementary units of the second group in the table. SI is now the preferred metric system, and many technical societies recommend its use. A comprehensive account of SI, its applications, and many conversion factors are contained in SMP (1982).

The following definitions apply to the base and supplementary SI units:*

Meter. The meter is the length equal to 1,650,763.73 wavelengths under

---

*Courtesy of the Institute of Electrical and Electronic Engineers, New York, N.Y.

TABLE B.1   BASE AND SUPPLEMENTARY UNITS OF THE INTERNATIONAL
SYSTEM OF UNITS (SI)

Quantity	Unit	Symbol	Dimension Symbol
	Base units		
Length	Meter	m	L
Mass	Kilogram	kg	M
Time	Second	sec	T
Electric current	Ampere	A	$M^{1/2}L^{1/2}T^{-1}$
Thermodynamic temperature	Kelvin	K	θ
Amount of substance	Mole	mol	M
Luminous intensity[a]	Candela	cd	ML^2T^{-3}
	Supplementary units		
Plane angle	Radian	rad	*b*
Solid angle	Steradian	sr	*b*

Source: After ASTM/IEEE, (1976).
[a]Dimensionally equal to radiant intensity in Table C.1.
[b]Suppressed.

vacuum of the radiation corresponding to the transition between the levels
$2p_{10}$ and $5d_5$ of the krypton-86 atom.

Kilogram. The kilogram is the mass corresponding to the mass of the
international prototype of the kilogram.

Second. The second is the duration of 9,192,631,770 periods of the radiation corresponding to the transition between the two hyperfine levels of the
ground state of the cesium-133 atom.

Ampere. The ampere is that constant current that, if maintained in two
straight parallel conductors of infinite length, of negligible circular cross
section, and placed 1 m apart under vacuum, would produce between these
conductors a force equal to 2×10^{-7} N m^{-1}

Kelvin. The kelvin, unit of the thermodynamic temperature, is the fraction 1/273.16 of the thermodynamic temperature of the triple point of
water.

Mole. The mole is the amount of substance of a system that contains as
many elementary entities as there are atoms in 0.012 kg of carbon-12.

Candela. The candela is the luminous intensity, in the perpendicular
direction, of a surface of 1/600,000 m^2 of blackbody at the temperature of
freezing platinum under a pressure of 101,325 Nm^{-2}.

Radian. The radian is the plane angle between two radii of a circle that
cut off on the circumference an arc equal to the length of the radius.

Steradian. The steradian is the solid angle that, having its vertex in the center of a sphere, cuts off an area of the surface of the sphere equal to that of a square with sides of length equal to the radius of the sphere.

Numerous additional units are derived from the SI base and supplementary units. The derived units of interest here are *force, energy,* and *power.* The SI unit of force is the *newton* N, defined as that force required to accelerate a 1 kg mass at 1 m sec^{-2}. In the CGS system the unit of force is the *dyne,* which is the force required to accelerate a 1 g mass at 1 cm sec^{-2}. The SI unit of energy is the *joule* J, which represents a force of 1 N acting through a distance of 1 m. In the CGS system the unit of energy is the *erg,* which represents a force of 1 dyn acting through a distance of 1 cm. Power is the time rate of expending energy or doing work, and the SI unit is the *watt,* defined as 1 J sec^{-1}. Conversely, 1 J = 1 W sec. The CGS unit of power is the erg sec^{-1}.

The SI units of force, energy, and power are the same regardless of whether the process is mechanical, electrical, chemical, or nuclear. A force of 1 N applied for a distance of 1 m can produce 1 J of heat, which is identical with what 1 W of electric power can produce in 1 sec.

In spectroscopy the wavenumber $\bar{\nu}$ in cm^{-1} is used, not only to specify a spectral location, but also as a measure of the energy itself. From Eq. (1.14) it is seen that the wavenumber of a photon is linearly related to its energy by Planck's constant h and the speed of light c, which act as porportionality constants. This photon we associate with a single transition, either absorptive or emissive, hence the energy is referred to a single molecule. Substituting for h and c in Eq. (1.14) their values from Appendix A we get

$$1 \text{ cm}^{-1} = 1.9863 \times 10^{-16} \text{ erg molecule}^{-1}$$

$$= 1.9863 \times 10^{-23} \text{ J molecule}^{-1} \qquad (B.1)$$

The number of 1 cm^{-1} units per joule then is

$$1 \text{ J} = 5.0345 \times 10^{22} \text{ cm}^{-1} \qquad (B.2)$$

This means that, when the spectral location of a line is expressed in cm^{-1} units, the energy of the corresponding transition is equal to that number of cm^{-1} units multiplied by the value of Eq. (B.1). It follows that, if the values of two energy levels are expressed in cm^{-1}, the difference between the values gives the wavenumber, thus the reciprocal of the wavelength, of the transition. This is the convenience of using the quantity cm^{-1} as an energy unit.

An energy unit often used in the physics of atoms and molecules is the electron volt (eV). One electron volt represents the kinetic energy acquired by an electron in being accelerated in an electric field through a potential

TABLE B.2 CONVERSION FACTORS FOR ENERGY UNITS

Multiply Number of ⟶ by ⟶ to Obtain

to Obtain	Joules	Ergs	Electron Volts	cm^{-1} molecule^{-1}	15°C g cal
Joules	1	10^{-7}	1.6021×10^{-19}	1.9863×10^{-23}	4.1855
Ergs	10^7	1	1.6021×10^{-12}	1.9863×10^{-16}	4.1855×10^7
Electron volts	6.2418×10^{18}	6.2418×10^{11}	1	1.2397×10^{-4}	2.6123×10^{19}
cm^{-1} molecule^{-1}	5.0345×10^{22}	5.0345×10^{15}	8.0665×10^3	1	2.1072×10^{23}
15°C g cal	0.2389	2.3890×10^{-6}	3.8280×10^{-20}	4.7457×10^{-24}	1

Source: Data from List (1966), USSA (1976), and Valley (1965).

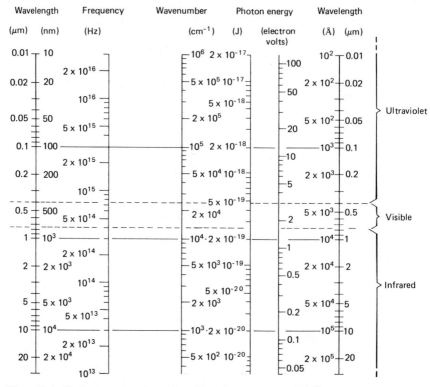

Figure B.1. Radiant energy conversion chart. Courtesy of Radio Corporation of America, Solid State Division, Electro-Optics and Devices, Lancaster, Pennsylvania.

difference of 1 V. The kinetic energy is acquired at the expense of the potential energy, which decreases according to

$$e \times \Delta V$$

where e is the charge in coulombs on an electron, and ΔV is the potential difference in volts. Since the electron charge is 1.6020×10^{-19} C, and the potential difference is 1 V, we have

$$1 \text{ eV} = 1.6020 \times 10^{-12} \text{ erg}$$

$$= 1.6020 \times 10^{-19} \text{ J} \tag{B.3}$$

The same result can be obtained from the laws of motion by considering the force on the electron and its resulting acceleration and velocity.

Table B.2 lists the conversion factors for the energy units discussed above. Figure B.1 presents a nomograph for quick conversions, with the μm scale at both margins to permit aligning a straight edge across the other scales.

Appendix C

Radiometric Quantities

Electromagnetic energy is inseparably joined to the fact of its propagation. Whether considered as waves having a continuous spatial function or as quanta distributed discontinuously in space, the energy is known *observably* only in its passing. Paraphrasing Skilling (1948): "An excess of electromagnetic energy in space cannot be confined, nor can it stand still, any more than a mound of water can be stationary on the surface of a lake." The time rate at which electromagnetic energy flows past a given point is the basic phenomena of all optics. The quantity corresponding to this rate is called *radiant flux* (from *fluere:* to flow), or *radiant power,* or simply *light.* We should keep in mind that for radiant energy to be detected or measured it usually must be absorbed and converted to some other form of energy—the radiant nature must be destroyed. The measuring of radiant energy and power within specified geometric and spectral limits is the discipline known as *radiometry.* No phenomena of optics need to be tied more closely to radiometry than do absorption and emission.

In this book it is assumed that readers are familiar generally with radiometric principles and measurement units. For reminder and reference we present here a list of the radiometric quantities along with several comments on the basics. Persons who wish to go beyond this summary are encouraged to consult among the many treatments in the literature, some of which are listed below. Entries marked with an asterisk are not for casual reading; they are quite detailed but will reward careful study.

ANSI (1967)	*Nicodemus (1976, 1979)
Drummond (1970)	*Nicodemus and Kostkowski (1977, 1978)
Engstrom (1974)	Stimson (1974)
Grum and Becherer (1979)	Williams and Becklund (1972)
McCartney (1976)	Wyatt (1977)
Nicodemus (1967, 1969)	Zissis (1979)

In dealing with radiometry this volume employs the International System of Units (SI), which has been derived from the MKSA (meter–kilogram–second–ampere) system. For practical purposes the SI radiometric quantities are the same as those in ANSI (1967). The SI units of energy and power are *joule* and *watt.* Alternative units in common use are

Reciprocal centimeter (cm^{-1})
Electron volt (eV)
Number of photons (numeric)

Appendix B gives numerical factors and a nomograph for conversions between the energy units in common use.

The principal radiometric quantities and their defining equations are listed in Table C.1. As written they refer to energy or flux over all wavelengths and could be qualified by the word *total*. They can be restricted to a narrow spectral range by prefacing the name with the word *spectral* and adding a subscript λ to the symbol. This subscript indicates that the quantity is differential with respect to wavelength, that is, the flux lies within a very small interval of wavelength centered at the stated wavelength. When the quantity is considered as a function of wavelength, a parenthetical λ is placed after the symbol.

A spectral quantity can be stated in terms of photons instead of joule or watt, and then a prime superscript is added to the symbol. The quantity so designated is spectral because photon energy is a function of wavelength, as discussed in Section 1.3.2. In such terms radiant energy is stated as number of photons at a specified wavelength, per unit wavelength interval. Radiant flux is stated the same way, but with the added stipulation *per second*.

A finite amount of flux requires a finite spectral width. An infinitely narrow interval represents *zero* flux, just as a conventional spectrometer having zero slit width could not transmit flux to the detector. The units for any spectral quantity must specify the wavelength interval, which is usually taken as equal to the wavelength unit itself. Energy over a spectral interval is found from

$$Q = \int_{\lambda_1}^{\lambda_2} Q(\lambda)\, d\lambda$$

which yields the total energy when the limits are extended to 0 and ∞. When the spectral distribution of flux is that of a blackbody, actual integration can be avoided by using tabulated functions available in the literature.

The bare phenomenology of radiant flux is noted briefly. Radiant density w is the electromagnetic energy content of a unit volume of space, as shown by the equation in Table C.1. Because this energy cannot be confined, it transports itself through the unit volume at the velocity of light. The product of energy density and velocity is equal to the time rate of energy flow per unit area, which can be visualized at either end face of the unit volume.

TABLE C.1 PRINCIPAL RADIOMETRIC QUANTITIES

Symbol	Quantity	Definition	Dimension	Common Unit
Q	Radiant energy	(Given)	ML^2T^{-2}	Joule (J)
w	Radiant density	$w = dQ/dV$	$ML^{-1}T^{-2}$	Joule per cubic meter (J m^{-3})
Φ	Radiant flux	$\Phi = dQ/dt$	ML^2T^{-3}	Watt (W) or joule per second (J sec^{-1})
	Radiant flux density at a surface			
M	Radiant exitance	$M = d\Pi/dA$	MT^{-3}	Watt per square meter (W m^{-2})
E	Irradiance	$E = d\Phi/dA$	MT^{-3}	Watt per square meter (W m^{-2})
I	Radiant intensity	$I = d\Phi/d\omega$	ML^2T^{-3}	Watt per steradian (W sr^{-1})
L	Radiance	$L = d^2\Phi/d\omega(dA \cos\theta$ $= dI/dA \cos\theta$	MT^{-3}	Watt per steradian per square meter (w sr^{-1} m^{-2})

Source: ANSI (1967). Courtesy of Illuminating Engineering Society.

Since the dimensions of energy density are $ML^{-1}T^{-2}$, and those of velocity are LT^{-1}, we have

$$ML^{-1}T^{-2} \times LT^{-1} = MT^{-3}$$

which are the dimensions of power per unit area. Thus the quantity *radiant power* or *flux* expresses the rate of energy flow across or through a surface, which may be either real or conceptual, such as the cross section of an optical beam. Radiant flux is the thing usually meant by the common use of the term *radiation,* which we restrict to mean the *process* of sending forth radiant energy.

Appendix D

Radiometric Properties of Matter

The gross properties of matter known as reflectance, absorptance, transmittance, and emissivity are basic factors in radiometry. These properties are dimensionless, and their defining equations are listed in Table D.1. Because each of these properties often varies strongly with wavelength, it is proper to specify whether a given value is *spectral,* for a narrow wavelength, or *total,* as averaged over a wavelength band. These properties are discussed to varying extents in most optics textbooks that deal with radiometry.

Looking now at the entries in Table D.1, *reflectance* ρ is defined as the ratio of reflected flux to incident flux. The subject of reflectance can be more involved than is implied by this definition. A detailed treatment is given by Nicodemus et al. (1977). Reflectance, however, does not enjoy a significant role in absorption–emission. *Absorptance* α is defined as the ratio of absorbed to incident flux, and must be distinguished from *absorption coefficient,* treated in Section 1.4.3. The flux that is neither reflected nor absorbed is transmitted. This property is expressed by the *transmittance* τ, defined as the ratio of transmitted to internally incident flux (external flux corrected for surface reflection where appropriate).

Thus the incident flux is disposed according to

$$\Phi_i = \Phi_i(\rho + \alpha + \tau) \tag{D.1}$$

Division by Φ_i gives

$$\rho + \alpha + \tau = 1 \tag{D.2}$$

TABLE D.1   RADIOMETRIC PROPERTIES OF MATTER

Symbol	Property	Defining Equation[a]	Range of Values
α	Absorptance	$\alpha = \Phi_\alpha/\Phi_i$	0–1
ρ	Reflectance	$\rho = \Phi_\rho/\Phi_i$	0–1
τ	Transmittance	$\tau = \Phi_\tau/\Phi_i$	0–1
ϵ	Emissivity	$\epsilon = M/M_{bb}$	0–1

Source: ANSI (1967). Courtesy of Illuminating Engineering Society.
[a]The subscript i denotes the incident radiant flux, and the subscripts α, ρ, τ denote that which is absorbed, reflected, or transmitted. The subscript bb refers to a blackbody whose emissivity is unity.

For a completely opaque material, the transmittance is zero, so that

$$\alpha = 1 - \rho \tag{D.3}$$

This says that a good absorber is a poor reflector, and vice versa, which is one form of Kirchhoff's law. In the atmosphere the reflectance of a parcel of air can be taken as zero if scattering is excluded. Then we have from Eq. (D.2),

$$\alpha = 1 - \tau \tag{D.4}$$

which has many applications in atmospheric optics.

The ability of a material to emit radiant energy is expressed by its *emissivity* ϵ. As shown by the final entry in Table D.1, this property is defined as the ratio of the radiant exitance M of a material to the radiant exitance M_{bb} of a blackbody at the same temperature. A blackbody is the ideal radiator whose emissivity is equal to unity; its concept and far-reaching laws are discussed in Section 1.5. Readers will recall that in customary radiometry the idea of emissivity is usually applied to the surfaces of liquids and solids. In the atmosphere, however, we are concerned with the emissivity (and absorptance) of gases *observed in depth*. Generally stated, such depth exists for a relatively short path in a high concentration of an absorbing gas and for a longer path in a lower concentration. The idea is quantitatively expressed by the *optical thickness* T defined by Eq. (1.31). As with the other entries in Table D.1, the qualification *spectral* or *total* applies to emissivity, with corresponding meanings.

Appendix **E**

Blackbody Spectral Radiance

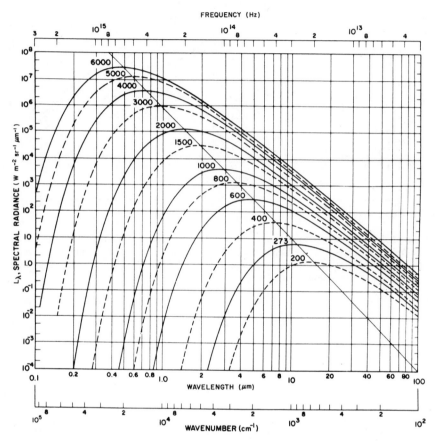

Figure E.1. Power spectral radiance of a blackbody, per unit wavelength interval, at the Kelvin temperature shown on each curve. From Valley (1965). *Handbook of Geophysics and Space Environments.* © 1965 by McGraw-Hill, Inc. Used with the permission of McGraw-Hill Book Co.

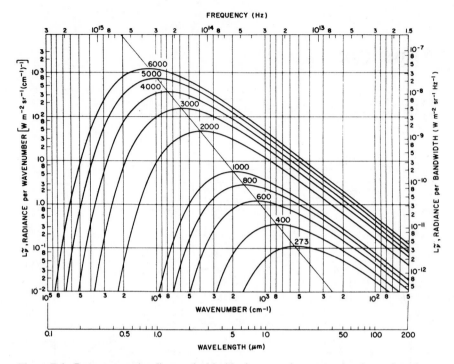

Figure E.2. Power spectral radiance of a blackbody, per unit wavenumber interval, and per unit frequency interval, at the Kelvin temperature shown on each curve. From Valley (1965). *Handbook of Geophysics and Space Environments.* © 1965 by McGraw-Hill, Inc. Used with the permission of McGraw-Hill Book Co.

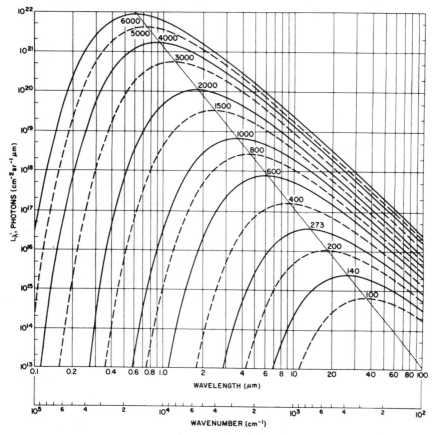

Figure E.3. Photon spectral radiance of a blackbody, per unit wavelength interval, at the Kelvin temperature shown on each curve. From Valley (1965). *Handbook of Geophysics and Space Environments.* © 1965 by McGraw-Hill, Inc. Used with the permission of McGraw-Hill Book Co.

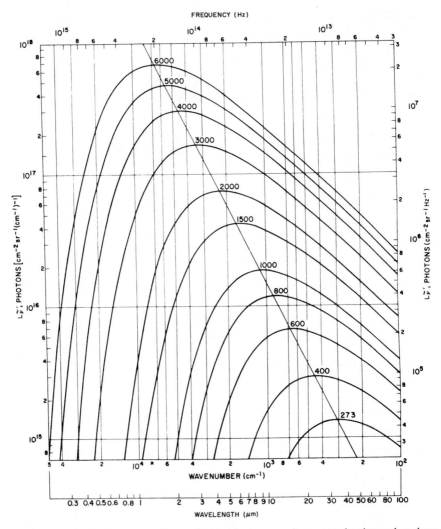

Figure E.4. Photon spectral radiance of a blackbody, per unit wavenumber interval, and per unit frequency interval, at the Kelvin temperature shown on each curve. From Valley (1965). *Handbook of Geophysics and Space Environments.* © 1965 by McGraw-Hill, Inc. Used with the permission of McGraw-Hill Book Co.

Appendix F

Integrated Fractional Exitance of a Blackbody

TABLE F.1   BLACKBODY INTEGRATED EXITANCE, $M_{0-\lambda}/M_{0-\infty}$, AS A FUNCTION OF THE PRODUCT λT (cm K)

cm K			cm K			cm K			cm K		
0.050	1.3652	9	0.095	1.7772	4	0.300	2.7454	1	0.700	8.0885	1
0.051	2.2642	9	0.096	2.0204	4	0.305	2.8585	1	0.720	8.1993	1
0.052	3.6788	9	0.097	2.2901	4	0.310	2.9712	1	0.740	8.3020	1
0.053	5.8629	9	0.098	2.5885	4	0.315	3.0833	1	0.760	8.3974	1
0.054	9.1749	9	0.099	2.9179	4	0.320	3.1947	1	0.780	8.4861	1
0.055	1.4113	8	0.100	3.2804	4	0.325	3.3053	1	0.800	8.5687	1
0.056	2.1358	8	0.105	5.6770	4	0.330	3.4150	1	0.820	8.6455	1
0.057	3.1829	8	0.110	9.2957	4	0.335	3.5237	1	0.840	8.7172	1
0.058	4.6745	8	0.115	1.4510	3	0.340	3.6314	1	0.860	8.7840	1
0.059	6.7710	8	0.120	2.1727	3	0.345	3.7379	1	0.880	8.8465	1
0.060	9.6798	8	0.125	3.1370	3	0.350	3.8432	1	0.900	8.9048	1
0.061	1.3667	7	0.130	4.3866	3	0.355	3.9474	1	0.920	8.9594	1
0.062	1.9069	7	0.135	5.9631	3	0.360	4.0502	1	0.940	9.0105	1
0.063	2.6307	7	0.140	7.9053	3	0.365	4.1517	1	0.960	9.0584	1
0.064	3.5907	7	0.145	1.0248	2	0.370	4.2518	1	0.980	9.1033	1
0.065	4.8510	7	0.150	1.3023	2	0.375	4.3506	1	1.00	9.1455	1
0.066	6.4902	7	0.155	1.6254	2	0.380	4.4479	1	1.05	9.2402	1
0.067	8.6028	7	0.160	1.9962	2	0.385	4.5438	1	1.10	9.3217	1
0.068	1.1302	6	0.165	2.4161	2	0.390	4.6382	1	1.15	9.3921	1
0.069	1.4723	6	0.170	2.8858	2	0.395	4.7312	1	1.20	9.4532	1
0.070	1.9025	6	0.175	3.4056	2	0.400	4.8227	1	1.25	9.5065	1
0.071	2.4393	6	0.180	3.9754	2	0.410	5.0012	1	1.30	9.5531	1
0.072	3.1045	6	0.185	4.5944	2	0.420	5.1738	1	1.35	9.5942	1
0.073	3.9230	6	0.190	5.2613	2	0.430	5.3404	1	1.40	9.6304	1
0.074	4.9236	6	0.195	5.9749	2	0.440	5.5012	1	1.45	9.6624	1
0.075	6.1392	6	0.200	6.7331	2	0.450	5.6563	1	1.50	9.6909	1
0.076	7.6070	6	0.205	7.5339	2	0.460	5.8057	1	1.55	9.7163	1
0.077	9.3692	6	0.210	8.3750	2	0.470	5.9495	1	1.60	9.7390	1
0.078	1.1473	5	0.215	9.2538	2	0.480	6.0880	1	1.65	9.7594	1
0.079	1.3971	5	0.220	1.0168	1	0.490	6.2212	1	1.70	9.7777	1
0.080	1.6923	5	0.225	1.1114	1	0.500	6.3494	1	1.75	9.7942	1

TABLE F.1   BLACKBODY INTEGRATED EXITANCE, $M_{0-\lambda}/M_{0-\infty}$, AS A
FUNCTION OF THE PRODUCT λT (cm K) (Continued)

cm K		cm K		cm K		cm K	
0.081	2.0393 5	0.230	1.2091 1	0.510	6.4727 1	1.80	9.8091 1
0.082	2.4453 5	0.235	1.3094 1	0.520	6.5912 1	1.85	9.8226 1
0.083	2.9183 5	0.240	1.4122 1	0.530	6.7051 1	1.90	9.8349 1
0.084	3.4668 5	0.245	1.5171 1	0.540	6.8146 1	1.95	9.8461 1
0.085	4.1002 5	0.250	1.6239 1	0.550	6.9198 1	2.00	9.8563 1
0.086	4.8287 5	0.255	1.7324 1	0.560	7.0209 1		
0.087	5.6633 5	0.260	1.8423 1	0.570	7.1182 1		
0.088	6.6159 5	0.265	1.9533 1	0.580	7.2116 1		
0.089	7.6993 5	0.270	2.0653 1	0.590	7.3014 1		
0.090	8.9269 5	0.275	2.1780 1	0.600	7.3877 1		
0.091	1.0314 4	0.280	2.2911 1	0.620	7.5505 1		
0.092	1.1874 4	0.285	2.4047 1	0.640	7.7010 1		
0.093	1.3626 4	0.290	2.5183 1	0.660	7.8402 1		
0.094	1.5586 4	0.295	2.6320 1	0.680	7.9691 1		

Note: The one digit number following the five digit number indicates the negative power of
10 by which the five digit entry is to be multiplied. From List (1966). Reprinted by permission
of the Smithsonian Institution Press from Smithsonian Meteorological Tables, 6th rev. ed.,
Smithsonian Institution, Washington, D.C. (1966).

Appendix   **G**

Transmittances of Three Sea-Level Paths

Curve	Path Length	Date	Time	Temp.	RH	Precipitable Water	Visual range
A	1000	3-20 - 56	3 PM	37 °F	62%	1.1 mm	22 mi
B	3.4 mi	3-20 - 56	10 PM	34.5 °F	47%	13.7 mm	16 mi
C	10.1 mi	3-21 - 56	12 AM	40.5 °F	48%	52.0 mm	24 mi

Window definitions			
I	0.72 - 0.94 μm	V	1.90 - 2.70 μm
II	0.94 - 1.13 μm	VI	2.70 - 4.30 μm
III	1.13 - 1.38 μm	VII	4.30 - 6.0 μm
IV	1.38 ÷ 1.90 μm	VIII	6.0 - 15.0 μm

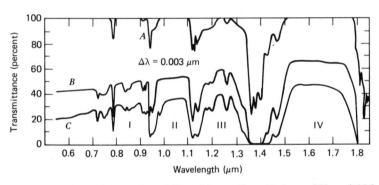

Figure G.1. Transmittances from 0.72 to 1.8 μm. From Taylor and Yates (1956).

285

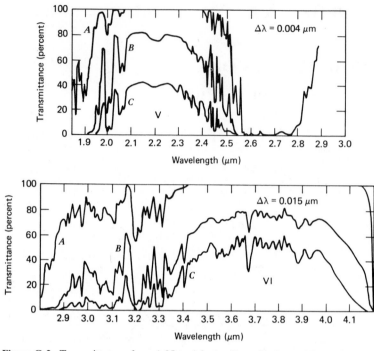

Figure G.2. Transmittances from 1.85 to 4.2 μm. From Taylor and Yates (1956).

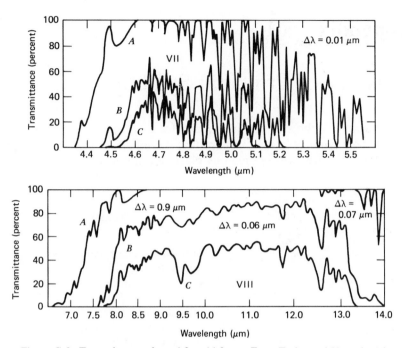

Figure G.3. Transmittances from 4.2 to 14.0 μm. From Taylor and Yates (1956).

Appendix **H**

Glossary of Principal Symbols

The principal symbols of physical quantities used in this book are listed below. Some symbols formed from principal symbols by adding subscripts, and others that are used in only one place, are not listed.

Symbol	First used in Section	Definition
$A(\nu)$	7.1.3	Spectral absorptance
$\bar{A}$	6.3.1	Mean fractional absorptance over a spectral interval
A_{nm}	6.2.1	Einstein coefficient for spontaneous emission
B	4.2.2	Rotational constant
B_{mn}	6.2.1	Einstein coefficient for absorption
B_{nm}	6.2.1	Einstein coefficient for induced emission
c	1.2.1	Speed of light
C_P	2.2.4	Specific heat at constant pressure
C_V	2.2.4	Specific heat at constant volume
e	4.1	Charge of an electron
e_s	2.1.2	Water vapor saturation pressure
e_v	2.1.2	Water vapor pressure
E	C	Irradiance
$\mathscr{E}$	1.3.1	Photon or molecular energy
$\mathscr{E}_J$	4.2.2	Energy of molecular rotational level
$\mathscr{E}_p$	3.1.3	Particle potential energy
$\mathscr{E}_v$	4.3.2	Energy of molecular vibrational level
$F(J)$	4.2.2	Rotational term value
g_i	3.2.2	Statistical weight of energy level
g	1.3.1	Acceleration of gravity
G_i	3.2.3	Density of energy states
$G(\nu)$	4.3.2	Vibrational term value
h	1.3.1	Planck's constant
I	C	Radiant intensity
I	4.2.1	Moment of inertia
j	6.1.1	Emission coefficient

Symbol	First used in Section	Definition
J	4.2.2	Rotational quantum number
$\mathbf{J}$	4.2.4	Total angular momentum vector
k	1.4.3	Linear absorption coefficient
K	1.4.3	Mass absorption coefficient
k	4.3.1	Vibrational force constant
K	4.4.3	Quantum number of $\mathbf{K}$
$\mathbf{K}$	4.4.3	Angular momentum vector
l	2.3.2	Molecular mean free path
L	C	Radiance
L	4.4.2	Quantum number of $\mathbf{L}$
$\mathbf{L}$	4.4.2	Angular momentum vector of electron orbital motion
L	1.3.1	Dimension of length
m	1.3.2	Mass of a photon or particle
m	6.2.1	Lower energy state of a transition
m'	4.2.1	Reduced mass of a molecule
M	1.5.1	Radiant exitance
M_L	4.4.2	Quantum number of axial component of orbital angular momentum
M_A	2.1.3	Molar mass
M	1.3.1	Dimension of mass
n	1.3.1	Quantum number (general)
n	6.2.1	Upper energy state of a transition
N	2.1.3	Molecular number density
$\mathbf{N}$	4.4.3	Angular momentum vector of nuclear rotation
N_A	2.1.3	Avogadro's number
N_L	2.1.3	Loschmidt's number
	1.3.2	Photon momentum
P	2.1.1	Pressure
$\mathcal{P}$	3.1.4	Probability
Q	C	Radiant energy
Q	2.2.4	Heat
r	4.2.1	Internuclear distance
R	2.2.1	Specific gas constant
R_*	2.2.1	Universal gas constant
R_{nm}	6.2.2	Matrix element of molecular dipole
S	2.2.5	Entropy
S	4.4.2	Quantum number of $\mathbf{S}$
$\mathbf{S}$	4.4.2	Angular momentum vector of electron spin
S	6.1.1	Spectral line strength
t	2.2.1	Temperature on the Celsius scale

Symbol	First used in Section	Definition
t	1.2.3	Time
T	1.5.2	Temperature on the Kelvin scale
T	1.3.1	Dimension of time
u	1.4.3	Mass of absorber
v	1.3.2	Velocity or speed
v	4.3.2	Vibrational quantum number
V	2.1.1	Volume
V_A	2.1.3	Molar volume
w	1.5.3	Radiant energy density
$\mathcal{w}$	3.3.1	Number of microstates in one energy level
$\mathcal{W}$	3.2.4	Thermodynamic probability of one macrostate
XYZ	2.4.1	Set of orthogonal coordinate axes
x	1.4.3	Distance along an x axis
y	2.4.1	Distance along a y axis
Z	2.1.1	Altitude
Z	3.3.5	Partition function
α	1.5.1	Absorptance
α	1.4.4	Spectral line half-width
α	3.3.3	Lagrange multiplier
β	3.3.3	Lagrange multiplier
γ	2.1.1	Temperature lapse rate
Γ	3.2.1	Total number of quantum states
δ	3.3.3	Small change of value
Δ	3.2.3	A differential or a small increase
Δ	4.4.2	An electronic state
ϵ	1.5.1	Emissivity
κ	1.5.2	Boltzmann's constant
λ	1.2.1	Electromagnetic wavelength
Λ	4.4.2	Quantum number of Λ
Λ	4.4.2	Angular momentum vector of axial component of electron orbital motion
μ	4.1	Electric dipole moment
ν	1.2.1	Electromagnetic frequency
$\tilde{\nu}$	1.2.1	Electromagnetic wavenumber
π	1.3.1	Ratio of circle circumference to diameter
Π	4.4.2	An electronic state
Π	3.3.1	Form the product of
ρ	D	Reflectance
ρ	1.4.3	Density of matter

Symbol	First used in Section	Definition
σ	1.5.2	Stefan–Boltzmann constant
σ	2.4.1	Standard deviation
σ	6.2.3	Molecular absorption cross section
Σ	4.4.2	An electronic state
Σ	4.4.2	Quantum number of Σ
$\boldsymbol{\Sigma}$	4.4.2	Axial component vector of electron spin
τ	2.3.2	Mean free time or time constant
τ	1.4.3	Transmittance
T	1.4.3	Optical thickness (depth)
ϕ	2.3.2	Mean collision rate
Φ	C	Radiant flux
Ψ	3.1.2	de Broglie wave function (time dependent)
ψ	3.1.3	de Broglie wave function (time independent)
ω	4.2.1	Angular velocity
Ω	3.2.4	Total number of microstates of a system
$\boldsymbol{\Omega}$	4.4.2	Angular momentum vector of total electronic motions

REFERENCES

Abels, L. L. and J. H. Shaw (1963). "Study of the total absorptance near 4.5u by two samples of N₂O as their total pressures and N₂O concentrations were independently varied." *J. Opt. Soc. Am.* **53,** 856–858.

Allen, H. C. and E. K. Plyler (1957). "ν_3 band of methane." *J. Chem. Phys.* **26,** 972.

Anderson, A., et al. (1967). "Self-broadening effects in the infrared bands of gases." *J. Opt. Soc. Am.* **57,** 240–246.

ANSI (1967). *American National Standard: Nomenclature and Definitions for Illuminating Engineering, Z7.1-1967.* Illuminating Engineering Society, New York.

AO (1969). *Appl. Opt.* **8,** 497–574. (Issue on "Interferometry vs. Grating Spectroscopy.")

AO (1978). *Appl. Opt.* **17,** 1315–1417. (Issue on "Fourier Transform Spectroscopy.")

Armstrong, B. H. (1967). "Spectrum line profiles: The Voigt function." *J. Quant. Spectrosc. Radiat. Transfer* **7,** 61–88.

Armstrong, B. H. and R. W. Nicholls (1972). *Emission, Absorption and Transfer of Radiation in Heated Atmospheres,* Pergamon Press, New York.

Arnold, C. B. and F. S. Simmons (1968). "Some measurements of sky emission in the 5–25 micrometers region." Report 8418-1-R. Infrared Physics Laboratory, Institute of Science and Technology, University of Michigan, Ann Arbor, AD 679587, NTIS, Springfield, Va.

Aronson, J. R., et al. (1975). "Tunable diode laser high resolution spectroscopic measurements of the ν_2 vibration of carbon dioxide." *Appl. Opt.* **14,** 1120–1127.

ASTM/IEEE (1976). *ASTM/IEEE Standard Metric Practice.* IEEE Std. 268-1976. Institute of Electrical and Electronic Engineers, New York.

Augason, G. C., et al. (1975). "Water vapor absorption spectra of the upper atmosphere (45–185 cm^{-1})." *Appl. Opt.* **14,** 2146–2150.

Babrov, H. J. and H. Casden (1968). "Strengths of forty-two lines in the ν_1 and ν_3 bands of water vapor." *J. Opt. Soc. Am.* **58,** 179–187.

Baker D., et al. (1981). "Development of infrared interferometry for upper atmospheric emission studies. *Appl. Opt.* **20,** 1734–1746.

Barrow, G. M. (1964). *The Structure of Molecules,* W. A. Benjamin, New York.

Bastin, J. A. (1966). "Extreme infrared atmospheric absorption." *Infrared Phys.* **6,** 209–221.

Bauman, R. P. (1962). *Absorption Spectroscopy,* Wiley, New York.

Beckman, J. E. and J. E. Harries (1975). "Submillimeter-wave atmospheric and astrophysical spectroscopy." *Appl. Opt.* **14,** 470–485.

Beiser, A. (1967). *Concepts of Modern Physics,* McGraw-Hill, New York.

Bell, E. E., et al. (1960). "Spectral radiance of sky and terrain at wavelengths between 1 and 20 microns. II. Sky measurements." *J. Opt. Soc. Am.* **50,** 1313–1320.

Bell, R. J. (1972). *Introductory Fourier Transform Spectroscopy,* Academic, New York.

Benedict, W. S. and R. F. Calfee (1967). "Line parameters for the 1.9 and 6.3 micron water vapor bands." ESSA Professional Paper 2, June 1967. GPO, Washington, D.C.

Benford, F. (1939). "Laws and corollaries of the blackbody." *J. Opt. Soc. Am.* **29,** 92–96.

Benford, F. (1943a). "The blackbody: Part I." *Gen. Electr. Rev.* July 1943, pp. 377–382.

Benford, F. (1943b). "The blackbody: Part II." *Gen. Electr. Rev.* August 1943, pp. 433–440.

Bennett, H. E., et al. (1960). "Distribution of infrared radiance over a clear sky." *J. Opt. Soc. Am.* **50,** 100–106.

Ben-Shalom, A., et al. (1981). "Long-path high-resolution atmospheric transmission measurements: Comparison with LOWTRAN 3B predictions: Comments." *Appl. Opt.* **20,** 171–172.

Bignell, K., et al. (1963). "On the atmospheric infrared continuum." *J. Opt. Soc. Am.* **53,** 466–479.

Bohr, N. (1949). "Discussion with Einstein on epistemological problems in atomic physics." *Library of Living Philosophers,* VII, pp. 201–241. Also in *World of the Atom,* H. A. Boorse and Lloyd Motz, eds., Basic Books, New York, 1966.

Born, M. (1951). *The Restless Universe,* Dover, New York.

Bose, S. N. (1924). "Planck's law and light quantum hypothesis." *Z. Phys.* **26.** Also in *World of the Atom,* H. A. Boorse and L. Motz, eds., Basic Books, New York (1966).

Bragg, W. (1940). *The Universe of Light,* Bell, London.

Bramson, M. A. (1968). *Infrared Radiation: A Handbook for Applications,* Plenum, New York.

Brinkman, R. T., et al. (1966). "Atomic and molecular species." In *The Middle Ultraviolet: Its Science and Technology,* A. E. Green, ed., Wiley, New York.

Burch, D. E. (1968). "Absorption of infrared radiant energy by CO_2 and H_2O. III. Absorption by H_2O between 0.5 and 36 cm^{-1} (278 μm–2 cm)." *J. Opt. Soc. Am.* **58,** 1383–1394.

Burch, D. E. and D. A. Gryvnak (1969). "Strengths, widths, and shapes of the oxygen lines near 13,100 cm^{-1} (7620 Å)." *Appl. Opt.* **8,** 1493–1499.

Burch, D. E. and D. A. Gryvnak (1971). "Absorption of infrared radiant energy by CO_2 and H_2O. V. Absorption by CO_2 between 100 and 1835 cm^{-1} (9.1–5.5 μm)." *J. Opt. Soc. Am.* **61,** 499–503.

Burch, D. E. and D. Williams (1962a). "Total absorptance by nitrous oxide bands in the infrared." *Appl. Opt.* **1,** 473–482.

Burch, D. E. and D. Williams (1962b). "Total absorptance of carbon monoxide and methane in the infrared." *Appl. Opt.* **1,** 587–594.

Burch, D. E., et al. (1962a). "Absorption line broadening in the infrared." *Appl. Opt.* **1,** 359–363.

Burch, D. E. et al. (1962b). "Total absorptance of carbon dioxide in the infrared." *Appl. Opt.* **1,** 759–765.

Burch, D. E., et al. (1963). "Total absorptance of water vapor in the near infrared." *Appl. Opt.* **2,** 585–589.

Burch, D. E., et al. (1968). "Absorption of infrared radiation by CO_2 and H_2O. II. Absorption by CO_2 between 8000 and 10,000 cm^{-1} (1–1.25 microns)." *J. Opt. Soc. Am.* **58,** 335–341.

Burch, D. E., et al. (1969). "Absorption of infrared radiant energy by CO_2 and H_2O. IV. Shapes of collision-broadened CO_2 lines." *J. Opt. Soc. Am.* **59,** 267–280.

Burch, D. E., et al. (1971). "Investigation of the absorption of infrared radiation by atmospheric gases: Water, nitrogen, nitrous oxide." Report AFCRL-71-0124. AFGL, Hanscom AFB, Bedford, Mass.

Button, K. J., ed. (1978). *Infrared and Submillimeter Waves,* Academic, New York.

Cantor, A. J. (1978). "Book review: *Optics of the Atmosphere,* by E. J. McCartney." *J. Quant. Elect.* **14,** 698–699.

Carli, B., et al. (1977). "Very-high-resolution far infrared measurements of atmospheric emission from aircraft." *J. Opt. Soc. Am.* **67,** 917–921.

Chai, An-Ti and D. Williams (1968). "Comparison of collision cross sections for line broadening in the CO fundamental." *J. Opt. Soc. Am.* **58,** 1395–1398.

Chamberlain, J. W. (1958). "Theories of the aurora," in *Advances in Geophysics,* Vol. 4, H. E. Landsberg and J. van Mieghem, eds. Academic, New York.

Chamberlain, J. W. (1961). *Physics of the Aurora and Airglow,* Academic, New York.

Chantry, G. W. (1971). *Submillimetre Spectroscopy,* Academic, New York.

Christy, R. W. and A. Pytte (1965). *The Structure of Matter: An Introduction to Modern Physics,* W. A. Benjamin, New York.

Clough, S. A., et al. (1977). "Algorithm for the calculation of absorption coefficient—Pressure broadened molecular transitions." Report, AFGL-TR-77-0164, Environmental Research Paper No. 607. AFGL, Hanscom AFB, Bedford, Mass.

Compton, A. H. (1923). "A quantum theory of the scattering of x-rays by light elements." *Phys. Rev.* **21,** 483–522. Also in *World of the Atom,* H. A. Boorse and L. Motz, eds. Basic Books, New York, 1966.

Connes, P. (1968). "How light is analyzed." *Sci. Am.,* Sept., 72–82.

Conrath, B. J. (1972). "Vertical resolution of temperature profiles obtained from remote radiation measurements." *J. Atmos. Sci.* **29,** 1262–1271.

Cook, G. R., et al. (1968). "Photoionization and absorption coefficients of N_2O." *J. Opt. Soc. Am.* **58,** 129–136.

Cornette, W. M. (1980). "Suggested modification to the total volume molecular scattering coefficient in LOWTRAN." *Appl. Opt.* **19,** A182.

Craig, R. A. (1965). *The Upper Atmosphere: Meteorology and Physics,* Academic, New York.

Curtis, A. R. and R. M. Goody (1956). "Thermal radiation in the upper atmosphere." *Proc. R. Soc. London Ser. A.* **236,** 193.

Davisson, C. J. (1928). "Are electrons waves?" *J. Franklin Inst.* **205,** 597–623. Also in *World of the Atom,* H. A. Boorse and L. Motz, eds., Basic Books, New York, 1966.

de Broglie, L. V. (1929). "The undulatory aspects of the electron." Nobel Prize Address, Stockholm, 1929. In *World of the Atom,* H. A. Boorse and L. Motz, eds., Basic Books, New York, 1966.

Deepak, A., ed. (1977). *Inversion Methods in Atmospheric Remote Sounding,* Academic, New York.

Ditchburn, R. W. (1958). *Light,* Blackie and Son, Glasgow.

Ditchburn, R. W. (1977). *Light,* 3d ed. Academic, New York.

Dix, M. G. (1976). "Blackbody radiant energy." Program No. 4000118, PPX-52. Professional Program Exchange. Texas Instruments, Lubbock, Texas.

Dotto. L. and H. Schiff (1978). *The Ozone War,* Doubleday, New York.

Draegert, D. A. and D. Williams (1968). "Collisional broadening of CO absorption lines by foreign gases." *J. Opt. Soc. Am.* **58,** 1399–1403.

Drayson, S. R. (1965). "Atmospheric slant path transmission in the 15μ CO_2 band." NASA Technical Note TN D-2744. Contract NAS-r-54(03), University of Michigan. N65-20944, NTIS, Springfield, Va.

Drayson, S. R. (1966). "Atmospheric transmission in the CO_2 bands between 12μ and 18μ." *Appl. Opt.* **5,** 385–391.

Drayson, S. R., et al. (1967). "The frequencies and intensities of carbon dioxide absorption lines between 12 and 18 microns." Technical Report, University of Michigan, Coll. Eng. Contract No. Cwb- 11376 with ESSA. PB 178 067, NTIS, Springfield, Va.

Drayson, S. R., et al. (1968). "Atmospheric absorption by carbon dioxide, water vapor, and oxygen." Final Report, University of Michigan, Coll. Eng. Contract No. Cwb-11376 with ESSA. PB 178 068, NTIS, Springfield, Va.

Drummond, A. J., ed. (1970). "Precision radiometry." In *Advances in Geophysics*, Vol. 14, H. E. Landsberg and J. van Mieghem, eds. Academic, New York.

Eddington, A. S. (1935). *New Pathways in Science*, Cambridge University Press, London. Also, University of Michigan Press, Ann Arbor, 1959.

Eddy, J. A., et al. (1969). "The far infrared transmission of the upper atmosphere." *J. Amos. Sci.* **26**, 1318–1328.

Einstein, A. (1917). "The quantum theory of radiation and atomic processes." *Physikalische Zeitschrift* **18**, 121–128. Also in *The World of the Atom*, H. A. Boorse and L. Motz, eds. Basic Books, New York (1966).

Eisberg, R. and R. Resnick (1974). *Quantum Physics of Atoms, Molecules, Solids, Nuclei, and Particles*, Wiley, New York.

Elder, T. and J. Strong (1953). "The infrared transmission of atmospheric windows." *J. Franklin Inst.* **255**, 189–208.

Eldridge, R. G. (1967). "Water vapor absorption of visible and near infrared radiation." *Appl. Opt.* **6**, 709–713.

Ellis, R. E. and B. Schurin (1969). "Integrated intensity measurements of carbon dioxide bands in the 4.82μ and 5.17μ regions." *Appl. Opt.* **8**, 2265–2268.

Elsasser, W. M. (1938). "Mean absorption coefficient and equivalent absorption coefficient of a band spectrum." *Phys. Rev.* **54**, 126–129.

Elsasser, W. M. (1942). *Heat Transfer by Infrared Radiation in the Atmosphere*, Harvard Meteorological Studies No. 6. Harvard University Press, Cambridge, Mass.

Elsasser, W. M. and M. F. Culbertson (1960). "Atmospheric radiation tables." *Meteor. Monographs* **4**, No. 23, AMS, Boston.

El-Sherbiny, M., et al. (1979). "High sensitivity point monitoring of ozone, and high resolution spectroscopy of the v_3 band of ozone." *Appl. Opt.* **18**, 1198–1203.

Eng, R. S., et al. (1977). "Tunable diode laser spectroscopic determination of v_3 band center of $^{14}C^{16}O_2$ at $4.5\mu m$." *Appl. Opt.* **16**, 3072–3074.

Engstrom, R. W., ed. (1974). *Electro-Optics Handbook*, RCA Commercial Engineering, Harrison, N.J.

Farrow, L. A., et al. (1979). "Nitric acid vapor line parameters measured by CO laser transmittance." *Appl. Opt.* **18**, 76–81.

FCST (1975). *Fluorocarbons and the Environment*. Report of Federal Task Force on Inadvertent Modification of the Stratosphere (IMOS). Federal Council for Science and Technology. GPO, Washington, D.C.

Fermi, E. (1937). *Thermodynamics*, Prentice-Hall, Englewood Cliffs, N.J.

Ferraro, J. R. and L. J. Basile, eds. (1978). *Fourier Transform Infrared Spectroscopy: Application to Chemical Systems*, Academic, New York.

Finkelnberg, W. (1947). "Conditions for blackbody radiations of gases." J. Opt. Soc. Am. **39**, 185–187.

Finkman, E., et al. (1967). "Integrated intensity of 3.3 μ band of methane." *J. Opt. Soc. Am.* **57,** 1130–1131.

Fleagle, R. G. and J. A. Businger (1963). *An Introduction to Atmospheric Physics,* Academic, New York.

Fleagle, R. G. and J. A. Businger (1980). *An Introduction to Atmospheric Physics,* 2nd ed., Academic, New York.

Frank, N. H. (1950). *Introduction to Electricity and Optics,* 2nd ed., McGraw-Hill, New York.

Gailey, T. D. (1969). "Optical absorption coefficient of molecular oxygen near 1215 angstroms." *J. Opt. Soc. Am.* **59,** 536–538.

Garbuny, M. (1965). *Optical Physics,* Academic, New York.

Gates, D. M. and W. J. Harrop (1963). "Infrared transmission of the atmosphere to solar radiation." *Appl. Opt.* **2,** 887–898.

Gates, D. M., et al. (1963). "Computed transmission spectra for 2.7-micron H_2O band." *Appl. Opt.* **2,** 1117–1121.

Gates, D. M., et al. (1964). "Line parameters and computed spectra for water vapor bands at 2.7 μm." National Bureau of Standards Monograph 71, August 1964. GPO, Washington, D.C.

Gebbie, H. A., et al. (1951). "Atmospheric transmission in the 1 to 14 μ region." *Proc. R. Soc. London Ser. A.* **206,** 87–107.

Gibson, G. A. and J. H. Pierluissi (1971). "Accurate formula for gaseous transmittance in the infrared." *Appl. Opt.* **10,** 1509–1518.

Giese, A. C. (1976). *Living With Our Sun's Ultraviolet Rays,* Plenum, New York.

Gille, J. C. (1968). "Inversion of radiometric measurements." *Bull. Am. Meteorol. Soc.* **49,** 903–912.

Goldberg, L. (1954). "The absorption spectrum of the atmosphere." In *The Earth as a Planet,* G. P. Kuiper, ed., University of Chicago Press, Chicago.

Goldberg, L. (1969). "Ultraviolet astronomy." *Sci. Am.* June 1969, 92–102.

Goldman, A. (1970). "Statistical band model parameters for long path atmospheric ozone in the 9–10 μ region." *Appl. Opt.* **9,** 2600–2604.

Goldman, A. and T. G. Kyle (1968). "A comparison between statistical model and line by line calculation with application to the 9.6 μ ozone and the 2.7 μ water vapor bands." *Appl. Opt.* **7,** 1167–1177.

Goldman, A., et al. (1970a). "Identification of the ν_3 NO_2 band in the solar spectrum observed from a balloon borne spectrometer." *Nature* **225,** 443–444.

Goldman, A., et al. (1970b). "Long path atmospheric ozone absorption in the 9–10 μ region observed from a balloon-borne spectrometer." *Appl. Opt.* **9,** 565–580.

Goldman, A., et al. (1970c). "Abundance of N_2O in the atmosphere between 4.5 and 13.5 km." *J. Opt. Soc. Am.* **60,** 1466–1468.

Goldman, A., et al. (1971). "Statistical band model parameters and integrated intensities for the 5.9, 7.5, and 11.3-μ bands of HNO_3 vapor." *Appl. Opt.* **10,** 65–73.

Goldman, A., et al. (1973a). "Balloon-borne infrared measurements of the vertical distribution of N_2O in the atmosphere." *J. Opt. Soc. Am.* **63,** 843–846.

Goldman, A., et al. (1973b). "Distribution of water vapor in the stratosphere as determined from balloon measurements of atmospheric emission spectra in the 24–29 μ region." *Appl. Opt.* **12,** 1045–1053.

Goldman, A., et al. (1973c). "Solar absorption in the CO fundamental region." *Astrophys. J.* **182**, 581–584.

Goldman, A., et al. (1978). *New Atlas of IR Solar Spectra,* preliminary ed. Department of Physics, University of Denver, Denver, Colo.

Goldman, A., et al. (1980). "High resolution IR balloon-borne solar spectra and laboratory spectra in the HNO_3 1720 cm^{-1} region." *Appl. Opt.* **19**, 3721–3724.

Goldman, A., et al. (1981). "Temperature dependence of HNO_3 absorption in the 11.3 μm region." *Appl. Opt.* **20**, 172–175.

Goncz, J. H. and P. B. Newell (1966). "Spectra of pulsed and continuous xenon discharges." *J. Opt. Soc. Am.* **56**, 87–92.

Goody, R. M. (1952). "A statistical model for water vapor absorption." *Quart. J. R. Meteor. Soc.* **78**, 165–169.

Goody, R. M. (1954). *The Physics of the Stratosphere,* Cambridge University Press, New York.

Goody, R. M. (1964). *Atmospheric Radiation: I. Theoretical Basis,* Oxford University Press, New York.

Gopal, E. S. (1974). *Statistical Mechanics and Properties of Matter: Theory and Applications,* Wiley, New York.

Gordon, M. A. and W. B. Burton (1979). "Carbon monoxide in the galaxy." *Sci. Am.,* May 1979, 54–67.

Gray, D. E., ed. (1957). *American Institute of Physics Handbook,* McGraw-Hill, New York.

Green, A. E. S., ed. (1966). *The Middle Ultraviolet: Its Science and Technology,* Wiley, New York.

Griffiths, P. R. (1975). *Chemical Infrared Fourier Transform Spectroscopy,* Wiley, New York.

Gringorten, I. I., et al. (1966). "Atmospheric humidity atlas—Northern hemisphere." Report AFCRL-66-621, Air Force Surveys in Geophysics No. 186. AFGL, Bedford, Mass.

Gruenzel, R. R. (1978). "Mathematical expressions for molecular absorption in LOWTRAN 3B." *Appl. Opt.* **17**, 2591–2593.

Grum, F. and R. Becherer (1979). *Optical Radiation Measurements, Vol. I: Radiometry,* Academic, New York.

Guillemin, V. (1968). *The Story of Quantum Mechanics,* Scribner, New York.

Harstad, K. G. (1972). "Rational approximation for the Voight line profile." *J. Opt. Soc. Am.* **62**, 827–828.

Haught, K. M. and D. M. Cordray (1978). "Long-path high-resolution atmospheric transmission measurements: Comparison with LOWTRAN 3B predictions." *Appl. Opt.* **17**, 2668–2670.

Haught, K. M. and D. M. Cordray (1979). "Long-path high-resolution atmospheric transmission measurements: Comparison with LOWTRAN 3B predictions. Author's reply to comments." *Appl. Opt.* **18**, 593.

Haught, K. M. and J. A. Dowling (1977). "Long-path high-resolution field measurements of absolute transmission in the 3.5 to 4.0 μm atmospheric window." *Opt. Lett.* **1**, 121–123.

Heisenberg, W. K. (1930). *Principles of the Quantum Theory,* University of Chicago Press, Chicago. Also in *World of the Atom,* H. A. Boorse and L. Motz, eds. Basic Books, New York, 1966.

Henderson, S. T. (1978). *Daylight and Its Spectrum,* 2d ed., Halsted Press (Wiley), New York.

Herriott, D. R. (1962). "Optical properties of a continuous He-Ne optical maser." *J. Opt. Soc. Am.* **52**, 31–37. Also in *Appl. Opt. Suppl.* **1**, 118–124.

Herschel, W. (1800a). "Investigation of the powers of the prismatic colours to heat and illuminate objects; with remarks." *Philos. Trans. R. Soc. London* **90**, 255.

Herschel, W. (1800b). "Experiments on the refrangibility of the invisible rays of the sun." *Philos. Trans. R. Soc., London* **90**, 284.

Herzberg, G. (1961). *Spectra of Diatomic Molecules,* Van Nostrand Reinhold, New York.

Herzberg, G. (1966). *Electronic Spectra of Polyatomic Molecules,* Van Nostrand Reinhold, New York.

Herzberg, G. (1971). *The Spectra and Structures of Simple Free Radicals,* Cornell University Press, Ithaca, N.Y.

Herzberg, G. and L. Herzberg (1957). "Constants of polyatomic molecules." In *American Institute of Physics Handbook,* D. E. Gray, ed., McGraw-Hill, New York.

HP (1982). "Blackbody thermal radiation." Program 00148D, for models HP 67/97.
"Blackbody thermal radiation." Program 00152C, for models HP 41C/41CV.
"Blackbody photon emission." Program 00910C, for models HP 41C/41CV.
Users' Library, Hewlett-Packard Co., Corvallis, Ore.

Hoover, G. M. and D. Williams (1969). "Infrared absorptance of carbon monoxide at low temperatures." *J. Opt. Soc. Am.* **59**, 28–33.

Hoover, G. M., et al. (1967). "Infrared absorption by overlapping bands of atmospheric gases." *Appl. Opt.* **6**, 481–487.

Howard, J. N., et al. (1956a). "Infrared transmission of synthetic atmospheres. II. Absorption by carbon dioxide." *J. Opt. Soc. Am.* **46**, 237–241.

Howard, J. N., et al. (1956b). "Infrared transmission of synthetic atmospheres. III. Absorption by water vapor." *J. Opt. Soc. Am.* **46**, 242–245.

Hsu, D. K., et al. (1978). *Spectral Atlas of Nitrogen Dioxide, 5530 to 6480 Å,* Academic, New York.

Huber, K. P. and G. Herzberg (1979). *Constants of Diatomic Molecules,* Van Nostrand Reinhold, New York.

Hudson, R. D., Jr. (1969). *Infrared System Engineering,* Wiley, New York.

Hudson, R. D. and V. L. Carter (1968). "Absorption of oxygen at elevated temperatures (300 to 900 K) in the Schumann–Runge system." *J. Opt. Soc. Am.* **58**, 1621–1629.

Huntley, H. E. (1958). *Dimensional Analysis,* Macdonald, London.

Huppi, E. R., et al. (1974). "Aircraft observations of the infrared emission of the atmosphere in the 700–2800 cm^{-1} region." *Appl. Opt.* **13**, 1466–1476.

Incropera, F. P. (1974). *Introduction to Molecular Structure and Thermodynamics,* Wiley, New York.

Inn, E. C. and Y. Tanaka (1953). "Absorption coefficient of ozone in the ultraviolet and visible regions." *J. Opt. Soc. Am.* **43**, 870–873.

Izatt, J. R., et al. (1969). "Positions, intensities, and widths of water-vapor lines between 475 and 692 cm^{-1}." *J. Opt. Soc. Am.* **59**, 19–27.

Jamieson, J. A., et al. (1963). *Infrared Physics and Engineering,* McGraw-Hill, New York.

Jancovici, B. (1973). *Statistical Physics and Thermodynamics,* Halsted Press (Wiley), New York.

Jansson, P. A. and C. L. Korb (1968). "A table of the equivalent widths of isolated lines with combined Doppler and collision broadened profiles." *J. Quant. Spectrosc. Radiat. Transfer* **8**, 1399–1409.

Javan, A., et al. (1961). "Population inversion and continuous optical maser oscillations in a gas discharge containing a He-Ne mixture." *Phys. Rev. Lett.* **6**, 106–110.

Jeans, J. H. (1959). *An Introduction to the Kinetic Theory of Gases*, Cambridge University Press, New York.

Jennings, R. E. and A. F. Moorwood (1971). "Atmospheric emission measurements with a balloon-borne Michelson interferometer." *Appl. Opt.* **10**, 2311–2318.

Junge, C. E. (1963). *Air Chemistry and Radioactivity*, Academic, New York.

Kaplan, L. D. (1954). "A quasi-statistical approach to the calculation of atmospheric transmission." *Proceedings of the Toronto Meteorological Conference, 1953*, pp. 43–48. AMS, Boston.

Kaplan, L. D. (1959). "Inference of atmospheric structure from remote radiation measurements." *J. Opt. Soc. Am.* **49**, 1004–1007.

Kaplan, L. D. and D. F. Eggers (1956). "Intensity and line-widths of the 15-micron band, determined by a curve of growth method." *J. Chem. Phys.* **25**, 876.

Kaplan, L. D., et al. (1956). "9.6-micron band of telluric ozone and its rotational analysis." *J. Chem. Phys.* **24**, 1183–1186.

Kennard, E. H. (1938). *Kinetic Theory of Gases*, McGraw Hill, New York.

Kielkopf, J. F. (1973). "New approximations to the Voigt function with applications to spectral-line profile analysis." *J. Opt. Soc. Am.* **63**, 987–995.

King, J. F. (1964). "Band absorption model for arbitrary line variance." *J. Quant. Spectrosc. Radiat. Transfer* **4**, 705–711.

King, J. F. (1965). "Modulated band absorption model." *J. Opt. Soc. Am.* **55**, 1498–1503.

Kingston, R. H. (1978). *Detection of Optical and Infrared Radiation*, Springer-Verlag, New York.

Kittel, C. (1958). *Elementary Statistical Physics*, Wiley, New York.

Kittel, C. (1969). *Thermal Physics*, Wiley, New York.

Kittel, C. and H. Kroemer (1980). *Thermal Physics*, 2d ed., W. H. Freeman, San Francisco.

Kneizys, F. X. (1978). "Atmospheric transmittance and radiance: The LOWTRAN code." *Proc. Soc. Photo-Opt. Instrum. Eng.* **142**, March.

Koepke, P. and H. Quenzel (1978). "Water vapor spectral transmission at wavelengths between 0.7 μm and 1 μm." *Appl. Opt.* **17**, 2114–2118.

Koller, L. R. (1965). *Ultraviolet Radiation*, 2nd ed., Wiley, New York.

Kondratyev, K. Y. (1969). *Radiation in the Atmosphere*, Academic, New York.

Kondratyev, K. Y., et al. (1964). "Balloon investigations of radiative fluxes in the free atmosphere." *Pure Appl. Geophys.* **58**, 187–203.

Kondratyev, K. Y., et al. (1965). "Atmospheric optics investigations on Mt. Elbrus." *Appl. Opt.* **4**, 1069–1076.

Kondratyev, V. (1965). *The Structure of Atoms and Molecules*, Dover, New York.

Krupenie, P. H. (1966). "The band spectrum of carbon monoxide." National Bureau of Standards, NSRDS–NBS5. July 8, 1966. GPO, Washington, D.C.

Kruse, P. W., et al. (1963). *Elements of Infrared Technology*, Wiley, New York.

Kuhn, W. R. and J. London (1969). "Infrared radiative cooling in the middle atmosphere." *J. Atmos. Sci.* **26**, 189–204.

Kyle, T. G. (1968). "Absorption by Doppler–Lorentz lines." *J. Quant. Spectrosc. Radiat. Transfer* **8**, 1455–1462.

Kyle, T. G. (1975). "Atlas of computed infrared atmospheric absorption spectra." Report NCAR-TN/STR-112. National Center for Atmospheric Research, Boulder, Colo.

Ladenberg, R. and F. Reiche (1913). "Ueber selektive absorption." *Ann. der Phys.* **42**, 181–203.

Lado-Bordowsky, O. and G. Amat (1979). "Laboratory and atmospheric measurements of HNO_3 to determine the amount of this constituent in the earth's atmosphere." *Appl. Opt.* **18**, 3400–3403.

Langer, R. M. (1957). Report on Signal Corps Contract DA-36-039-SC-72 351 (May 1957).

La Rocca, A. J. and R. E. Turner (1975). *Atmospheric Transmittance and Radiance: Methods of Calculation*, Environmental Research Institute of Michigan, University of Michigan, Ann Arbor.

Levi, L. (1968). *Applied Optics: A Guide to Modern Optical System Design*, Wiley, New York.

Levi, L., ed. (1974). *Handbook of Tables of Functions for Applied Optics*, CRC Press, Boca Raton, Fla.

List, R. J., ed. (1966). *Smithsonian Meteorological Tables*, 6th ed., Smithsonian Institution, Washington, D.C.

Loeb, L. B. (1934). *The Kinetic Theory of Gases*, McGraw-Hill, New York. Also, Dover, New York, 1961.

Long, D. A. (1977). *Raman Spectroscopy*, McGraw-Hill, New York.

Lucretius (c60 BC). *De Rerum Natura*. Classics Club, W. J. Black, Roslyn, N.Y.

Ludwig, C. B., et al. (1973). *Handbook of Infrared Radiation from Combustion Gases*. NASA SP-3080. Scientific and Technical Information Office, NASA, Washington, D.C.

Mandl, F. (1980). *Statistical Physics*, Wiley, New York.

Mantz, A. W., et al. (1976). "Emission spectrum of nitric oxide between 5 μm and 7 μm." *Appl. Opt.* **15**, 599–600.

Margolis, J. S., et al. (1974). "Abundance and rotational temperature of telluric methane as determined from the $2\nu_3$ band." *J. Atmos. Sci.* **31**, 823–827.

Maxwell, J. C. (1860). "Illustrations of the dynamical theory of gases." *Philos. Mag.* **19**, 19–32. Also in *World of the Atom*, H. A. Boorse and L. Motz, eds., Basic Books, New York, 1966.

McCartney, E. J. (1976). *Optics of the Atmosphere: Scattering by Molecules and Particles*, Wiley, New York.

McClatchey, R. A. and F. X. Kneizys (1979). "Long-path high-resolution atmospheric transmission measurements: Comparison with LOWTRAN 3B predictions; comments." *Appl. Opt.* **18**, 592–593.

McClatchey, R. A., et al. (1972). "Optical properties of the atmosphere, 3d ed." AFCRL Report 72-0497. AFGL, Hanscom AFB, Bedford, Mass.

McClatchey, R. A., et al. (1973). "AFCRL atmospheric absorption line parameters compilation." Report AFCRL-TR-73-0096, Environmental Research Papers No. 434. AFGL, Hanscom AFB, Bedford, Mass.

McCoy, J. H., et al. (1969). "Water vapor continuum absorption of carbon dioxide laser radiation near 10 μ." *Appl. Opt.* **8**, 1471–1478.

McFee, R. H. (1976a). *Blackbody Radiance*, Program No. 740005, PPX-52. Professional Program Exchange. Texas Instruments, Lubbock, Texas.

McFee, R. H. (1976b). *Blackbody Photon Radiance,* Program No. 740011, PPX-52. Professional Program Exchange. Texas Instruments, Lubbock, Texas.

Meyer-Arendt, J. R. (1972). *Classical and Modern Optics,* Prentice-Hall, Englewood Cliffs, N.J.

Minnaert, M. (1940). *The Nature of Light and Color in the Open Air,* Bell, London.

Monchalin, J. P., et al. (1977). "Determination of the speed of speed of light by absolute wavelength measurements of the R(14) line of the CO_2 9.4 μm band and the known frequency of this line." *Opt. Lett.* **I,** 5–7.

Moon, P. (1936). *The Scientific Basis of Illuminating Engineering,* McGraw-Hill, New York. Also, Dover, New York, 1961.

Murcray, D. G., et al. (1960). "Atmospheric absorptions in the infrared at high altitudes." *J. Opt. Soc. Am.* **50,** 107–112.

Murcray, D. G., et al. (1961). "Study of the 1.4-μ, 1.9-μ, and 6.3-μ water vapor bands at high altitudes." *J. Opt. Soc. Am.* **51,** 186–195.

Murcray, D. G., et al. (1962). "Optical measurements from high altitude balloons." *Appl. Opt.* **1,** 121–128.

Murcray, D. G., et al. (1965). "Comparison of experimental and theoretical slant path absorptions in the region from 1400 to 2500 cm^{-1}." *J. Opt. Soc. Am.* **55,** 1239–1246.

Murcray, D. G., et al. (1969a). "Presence of HNO_3 in the upper atmosphere." *J. Opt. Soc. Am.* **59,** 1131–1134.

Murcray, D. G., et al. (1969b). "Variation of the infrared solar spectrum between 700 cm^{-1} and 2240 cm^{-1} with altitude." *Appl. Opt.* **8,** 2519–2536.

NAS (1975). *Environmental Impact of Stratospheric Flight,* National Academy of Sciences, Washington, D.C.

NAS (1976a). *Halocarbons: Effects on Stratospheric Ozone,* National Academy of Sciences, Washington, D.C.

NAS (1976b). *Halocarbons: Environmental Effects of Chlorofluoromethane Release,* National Academy of Sciences, Washington, D.C.

Nelson, R. D., Jr. (1967). "Selected values of electric dipole moments for molecules in the gas phase." National Bureau of Standards, NSRDS-NBS10. GPO, Washington, D.C.

Nicodemus, F. E. (1967). "Radiometry." In *Applied Optics and Optical Engineering,* Vol. 4, R. Kingslake, ed., Academic, New York.

Nicodemus, F. E. (1969). "Optical resource letter on radiometry." *J. Opt. Soc. Am.* **59,** 243–248. Also *Am. J. Phys.* **38,** 43–50.

Nicodemus, F. E., ed. (1976). "Self-study manual on optical radiation measurements, Part I: Concepts." Chapters 1, 2, and 3. National Bureau of Standards Tech. Note 910-1. GPO, Washington, D.C.

Nicodemus, F. E., ed. (1979). "Self-study manual on optical radiation measurements: Part I: Concepts." Chapters 7, 8, and 9. National Bureau of Standards Tech. Note 910-4. GPO, Washington, D.C.

Nicodemus, F. E. and H. J. Kostkowski, eds. (1977). "Self-study manual on optical radiation measurements, Part I: Concepts." Chapters 4 and 5. National Bureau of Standards Tech Note 910-2. GPO, Washington, D.C.

Nicodemus, F. E. and H. J. Kostkowski, eds. (1978). "Self-study manual on optical radiation measurements, Part I: Concepts." Chapter 6. National Bureau of Standards Tech Note 910-3. GPO, Washington, D.C.

Nicodemus, F. E., et al. (1977). "Geometrical considerations and nomenclature for reflectance." National Bureau of Standards, Monograph 160. Oct. 1977. GPO, Washington, D.C.

Nolt, I. G., et al. (1971). "Far infrared absorption of the atmosphere above 4.2 km." *J. Atmos. Sci.* **28**, 238–241.

Nordstrom, R. J., et al. (1978). "Effects of oxygen addition on pressure-broadened water-vapor absorption in the 10-μm region." *Appl. Opt.* **17**, 2724–2729.

Osgood, T. H., et al. (1964). *Atoms, Radiation, and Nuclei,* Wiley, New York.

O'Shea, D. C. and L. G. Dodge (1974). "NO_2 concentration measurements in an urban atmosphere using differential absorption techniques." *Appl. Opt.* **13**, 1481–1486.

Palmer, C. H., Jr. (1957a). "Long path water vapor spectra with pressure broadening. I. 20 μ to 31.7 μ. *J. Opt. Soc. Am.* **47**, 1024–1028.

Palmer, C. H., Jr. (1957b). "Long path water vapor spectra with pressure broadening. II. 29 μ to 40 μ. *J. Opt. Soc. Am.* **47**, 1028–1031.

Palmer, C. H., Jr. (1960). "Experimental transmission functions for the pure rotation band of water vapor." *J. Opt. Soc. Am.* **50**, 1232–1242.

Parrish, J. A., et al. (1978). *UV-A: Biological Effects of Ultraviolet Radiation,* Plenum, New York.

Passman, S. and L. Larmore (1956). "Atmospheric transmission." Rand Paper P-897. Rand Corporation, Santa Monica, Calif.

Pauling, L. and E. B. Wilson (1935). *Introduction to Quantum Mechanics,* McGraw-Hill, New York.

Penner, S. S. (1959). *Quantitative Molecular Spectroscopy and Gas Emissivities,* Addison-Wesley, Reading, Mass.

Peterson, J. C., et al. (1979). "Water vapor–nitrogen absorption at CO_2 laser frequencies." *Appl. Opt.* **18**, 834–841.

Pierce, J. R. (1961). *An Introduction to Information Theory: Symbols, Signals, and Noise,* Harper, New York. Also, Dover, New York, 1980.

Pierluissi, J. H., et al. (1979). "Analysis of the LOWTRAN transmission functions." *Appl. Opt.* **18**, 1607–1612.

Pivovonsky, M. and M. R. Nagel (1961). *Tables of Blackbody Radiation Functions,* Macmillan, New York.

Planck, M. (1914). *The Theory of Heat Radiation,* P. Blakiston Son, London. Also, Dover, New York, 1959.

Plass, G. N. (1958). "Models for spectral band absorption." *J. Opt. Soc. Am.* **48**, 690–703.

Plass, G. N. (1960). "Useful representations for measurements of spectral band absorption." *J. Opt. Soc. Am.* **50**, 868–875.

Plass, G. N. and D. I. Fivel (1953). "Influence of Doppler effect and damping on line-absorption coefficient and atmospheric radiation transfer." *Astrophys. J.* **117**, 225–233.

Plyler, E. K., et al. (1955). "Velocity of light from the molecular constants of carbon monoxide." *J. Opt. Soc. Am.* **45**, 102–106.

Posener, D. W. (1959). "The shapes of spectral lines: Tables of the Voigt profile." *Australian J. Phys.* **12**, 184–196.

Rayleigh, Lord (1877). *The Theory of Sound,* Macmillan, London. Also, Dover, New York, 1945.

Rayleigh, Lord (1879). "Investigations in optics, with special reference to the spectroscope," *Philos. Mag.* **8**, 261–274, 403–411, 477–486. Also in *Scientific Papers of Lord Rayleigh,* Vol. I, pp. 415–428, Dover, New York, 1964.

Rayleigh, Lord (1889a). "The history of the doctrine of radiant energy," *Philos. Mag.* **27**, 265–270. Also in *Scientific Papers of Lord Rayleigh,* Vol. III, pp. 238–243, Dover, New York, 1964.

Rayleigh, Lord (1889b). "On the limit to interference when light is radiated from moving molecules." *Philos. Mag.* **27**, 298–304. Also in *Scientific Papers of Lord Rayleigh,* Vol. III, pp. 258–263, Dover, New York, 1964.

Rhine, P. E., et al. (1969). "Nitric acid vapor above 19 km in the earth's atmosphere." *Appl. Opt.* **8**, 1500–1501.

Richton, R. E. (1976). "NO line parameters measured by CO laser transmittance." *Appl. Opt.* **15**, 1686–1687.

Roberts, R. E., et al. (1976). "Infrared continuum absorption by atmospheric water vapor in the 8–12 μm window." *Appl. Opt.* **15**, 2085–2090.

Robinson, N., ed. (1966). *Solar Radiation,* Elsevier, New York.

Rodgers, C. D. (1976). "Approximate methods of calculating transmission by bands of spectral lines." Technical Note NCAR/TN-116+1A, March 1976. NCAR, Boulder, Colo.

Roney, P. L., et al. (1978). "Carbon dioxide spectral line frequencies for the 4.3 μm region." *Appl. Opt.* **17**, 2599–2604.

Rosenberg, R. M. (1977). *Principles of Physical Chemistry,* Oxford University Press, New York.

Rossi, B. (1965). *Optics,* Addison-Wesley, Reading, Mass.

Rothman, L. S. (1978a). "High-resolution atmospheric transmittance radiance: HITRAN and the data compilation." *Proc. Soc. Photo-Opt. Instrum. Eng.* **142**, March 1978.

Rothman, L. S. (1978b). "Update of the AFGL absorption line parameters compilation." *Appl. Opt.* **17**, 3517–3518.

Rothman, L. S. (1981). "AFGL atmospheric absorption line parameters: 1980 version." *Appl. Opt.* **20**, 791–795.

Rothman, L. S. and R. A. McClatchey (1977). "Updating of the AFCRL atmospheric absorption line parameters compilation." *Appl. Opt.* **15**, 2616.

Rothman, L. S., et al. (1978). "AFGL trace gas compilation." *Appl. Opt.* **17**, 507.

SA (1968). *Sci. Am.,* Sept. (Issue on "Light.")

Schmidt, S. C., et al. (1974). "Quantitative studies of HNO₃ vapor absorption in the 1700-2636 Å wavelength region." *Appl. Opt.* **13**, 1202–1208.

Schurin, B. D. and R. E. Ellis (1968). "Integrated intensity measurements of carbon dioxide in the 2.0 μ, 1.6 μ, and 1.43 μ regions." *Appl. Opt.* **7**, 467–470.

Sears, F. W. and G. L. Salinger (1975). *Thermodynamics, Kinetic Theory, and Statistical Thermodynamics,* 3d ed., Addison-Wesley, Reading, Mass.

Seeley, J. S. and J. T. Houghton (1961). "Spectroscopic observations of the vertical distribution of some minor constituents of the atmosphere." *Infrared Phys.* **1**, 116–132.

Seitz, W. S. and D. V. Lundholm (1964). "Elsasser model for band absorption: series representation of a useful integral." *J. Opt. Soc. Am.* **54**, 315–318.

Selby, J. E. and R. A. McClatchey (1972). "Atmospheric transmittance from 0.25 to 28.5 μm: Computer code LOWTRAN 2." AFCRL-72-0745. AFGL, Hanscom AFB, Bedford, Mass.

Selby, J. E. and R. A. McClatchey (1975). "Atmospheric transmittance from 0.25 to 28.5 μm: Computer code LOWTRAN 3." AFCRL-TR-75-0255. AFGL, Hanscom AFB, Bedford, Mass.

Selby, J. E., et al. (1976). "Atmospheric transmittance from 0.25 to 28.5 μm: Supplement LOWTRAN 3B." AFCRL-TR-76-0258. AFGL, Hanscom AFB, Bedford, Mass.

Selby, J. E., et al. (1978). "Atmospheric transmittance/radiance: Computer code LOWTRAN 4." AFGL-TR-78-0053. AFGL, Hanscom AFB, Bedford, Mass.

Shafer, J. H. and C. Young (1976). "Absolute integrated intensity for the 3.44 μm NO_2 band." *Appl. Opt.* **15**, 2551–2553.

Shaw, J. H. and W. L. France (1956). "Intensities and widths of single lines of the 4.7 micron CO fundamental." Scientific Report 4 on Project 587, Ohio State University Research Foundation, Columbus, Ohio.

Shaw, J. H. and J. T. Houghton (1964). "Spectral band absorptance of CO near 4.7 μm." *Appl. Opt.* **3**, 773–779.

Shettle, E. P., et al. (1980). "Suggested modification to the total volume molecular scattering coefficient in LOWTRAN: Comment." *Appl. Opt.* **19**, 2873–2874.

Shimanouchi, T. (1967a). "Tables of Molecular Vibrational Frequencies, Part 1." National Bureau of Standards, NSRDS-NBS6. March 1967. GPO, Washington, D.C.

Shimanouchi, T. (1967b). "Tables of Molecular Vibrational Frequencies, Part 2." National Bureau of Standards, NSRDS-NBS11. Oct. 1967. GPO, Washington, D.C.

Shimanouchi, T. (1968). "Tables of Molecular Vibrational Frequencies, Part 3." National Bureau of Standards, NSRDS-NBS17. March 1968. GPO, Washington, D.C.

Shumate, M. S., et al. (1976). "Water vapor absorption of carbon dioxide laser radiation." *Appl. Opt.* **15**, 2480–2488.

Skilling, H. H. (1948). *Fundamentals of Electric Waves,* Wiley, New York.

Sloan, R., et al. (1955). "Infrared emission spectrum of the atmosphere." *J. Opt. Soc. Am.* **45**, 455–460.

Sloan, R., et al. (1956). "Thermal radiation from the atmosphere." *J. Opt. Soc. Am.* **46**, 543–547.

Smith, F. G. and J. H. Thomson (1971). *Optics,* Wiley, New York.

Smith, R. A., et al. (1957). *The Detection and Measurement of Infra-Red Radiation,* Oxford University Press, New York.

SMP (1982). *Standard for Metric Practice.* IEEE Std. 268–1982. Institute of Electrical and Electronics Engineers, New York.

Sommerfeld, A. (1930). *Wave Mechanics,* Methuen, London.

Sommerfeld, A. (1964). *Thermodynamics and Statistical Mechanics: Lectures on Theoretical Physics,* Vol. V, Academic, New York.

Sproull, R. L. (1963). *Modern Physics,* 2d ed., Wiley, New York.

Stauffer, F. R. and T. E. Walsh (1966). "Transmittance of water vapor 14 to 20 microns." *J. Opt. Soc. Am.* **56**, 401–405.

Stern, O. and W. Gerlach (1922). "Experimental proof of space quantization." *Z. Phys.* **9**, 349–352. Also in *World of the Atom,* H. A. Boorse and L. Motz, eds., Basic Books, New York, 1966.

Stimson, A. (1974). *Photometry and Radiometry for Engineers,* Wiley, New York.

Stone, J. M. (1963). *Radiation and Optics,* McGraw-Hill, New York.

Streete, J. L. (1968). "Infrared measurements of atmospheric transmission at sea level." *Appl. Opt.* **7**, 1545–1549.

Streete, J. L., et al. (1967). "Near infrared atmospheric absorption over a 25-km horizontal path at sea level." *Appl. Opt.* **6**, 489–496.

Strong, J. (1941). "On a new method of measuring the mean height of the ozone in the atmosphere." *J. Franklin Inst.* **231**, 121–155.

Stull, V. R., et al. (1962). "Quasi-random model of band absorption." *J. Opt. Soc. Am.* **52**, 1209–1217.

Stull, R., et al. (1964). "The infrared transmittance of carbon dioxide." *Appl. Opt.* **3**, 243–254.

Sugden, T. M. and T. F. West, eds. (1980). *Chlorofluorcarbons in the Environment,* Halsted Press (Wiley), New York.

Tajime, T., et al. (1978). "Absorption characteristics of the γ-0 band of nitric oxide." *Appl. Opt.* **17**, 1290–1294.

Takeuchi, N., et al. (1978). "Detectivity estimation of the DAS lidar for NO_2." *Appl Opt.* **17**, 2734–2738.

Tanaka, Y., et al. (1953). "Absorption coefficients of gases in the vacuum ultraviolet. Part IV. Ozone." *J. Chem. Phys.* **21**, 1651–1653.

Taylor, J. H. and H. W. Yates (1956). "Atmospheric transmission in the infrared." NRL Rpt. 4759, July 2, 1956. NRL, Washington, D.C. Also in *J. Opt. Soc. Am.* **47**, 223.

TI (1978). "Programmable TI58/59 Specialty Packettes for Blackbody Radiation." Texas Instruments, Lubbock, Texas.

Tidwell, E. D., et al. (1960). Vibration–rotation bands of N_2O." *J. Opt. Soc. Am.* **50**, 1243–1263.

Tiwari, S. N. (1973). "Appropriate line profiles for radiation modeling in the detection of atmospheric pollutants." Report TR-73-T3. School of Engineering, Old Dominion University, Norfolk, Va.

Tiwari, S. N. (1978). "Models for Infrared Atmospheric Radiation." In *Advances in Geophysics,* Vol. 20, B. Saltzman, ed., Academic, New York.

Tolman, R. C. (1938). *The Principles of Statistical Mechanics,* Oxford University Press, New York. Also, Dover, New York, 1979.

Tomasi, C., et al. (1974). "A search for the effect in the atmospheric water vapor continuum." *J. Atmos. Sci.* **31**, 255–260.

Townes, C. H. and A. L. Schalow (1955). *Microwave Spectroscopy,* McGraw-Hill, New York. Also, Dover, New York, 1975.

Traub, W. A. and M. T. Stier (1976). "Theoretical atmospheric transmission in the mid- and far-infrared at four altitudes." *Appl. Opt.* **15**, 364–377.

Tubbs, L. D. and D. Williams (1972a). "Broadening of infrared absorption lines at reduced temperatures: Carbon dioxide." *J. Opt. Soc. Am.* **62**, 284–289.

Tubbs, L. D. and D. Williams (1972b). "Broadening of infrared absorption lines at reduced temperatures: Carbon monoxide in an atmosphere of carbon dioxide." *J. Opt. Soc. Am.* **62**, 423–427.

Tubbs, L. D. and D. Williams (1972c). "Foreign-gas broadening of nitrous oxide absorption lines." *Appl. Opt.* **11**, 551–553.

Tverskoi, P. N. (1965). *Physics of the Atmosphere: A Course in Meteorology,* NASA Technical Translation, NASA TTF-288. NTIS, Springfield, Va.

USSA (1962). *U.S. Standard Atmosphere, 1962*, GPO, Washington, D.C.

USSA (1976). *U.S. Standard Atmosphere, 1976*, NOAA-S/T 76-1562. GPO, Washington, D.C.

USSAS (1966). *U.S. Standard Atmosphere Supplements, 1966*, GPO, Washington, D.C.

Valley, S. L., ed. (1965). *Handbook of Geophysics and Space Environments*, McGraw-Hill, New York.

Vanasse, G. A., ed. (1977). *Spectrometric Techniques*, Vol. 1, Academic, New York.

Vanasse, G. A., ed. (1981). *Spectrometric Techniques*, Vol. II, Academic, New York.

Vanderwerf, D. F. and J. H. Shaw (1965). "Temperature dependence of the total absorptance of bands of CO and CH_4." *Appl. Opt.* **4**, 209–213.

Van Wylen, G. J. and R. E. Sonntag (1976). *Fundamentals of Classical Thermodynamics*, 2nd ed., Wiley, New York.

Vigroux, E. (1953). "Contribution à l'étude expérimentale de l'absorption de l'ozone." *Ann. Phys.* **8**, 709.

Vincenti, W. G. and C. H. Kruger, Jr. (1965). *Introduction to Physical Gas Dynamics*, Wiley, New York.

von Hippel, A. (1954). *Dielectrics and Waves*, Wiley, New York.

Wald, G. (1959). "Life and Light." *Sci. Am.* October, 92–108.

Walker, S. and H. Straw (1967). *Spectroscopy, Vol. 2: Ultraviolet, Visible, Infrared, and Raman Spectroscopy*, Chapman and Hall, London.

Walshaw, C. D. (1957). "Integrated absorption by the 9.6-μ band of ozone." *Quart. J. R. Meteor. Soc.* **83**, 315–321.

Wang, J. Y. (1978). "Temperature effect on the atmospheric transmission function in the 15-μm region." *Opt. Lett.* **2**, 169–171.

Wark, D. Q. (1970). "SIRS: An experiment to measure the free air temperature from a satellite." *Appl. Opt.* **9**, 1761–1766.

Wark, D. Q. and D. M. Mercer (1965). "Absorption in the atmosphere by the oxygen 'A' band." *Appl. Opt.* **4**, 839–844.

Wark, D. Q. and M. Wolk (1960). "An extension of a table of absorption for Elsasser bands." *Mon. Weather Rev.* **88**, 249–250.

Wäser, J. (1966). *Basic Chemical Thermodynamics*, Benjamin, New York.

Watanabe, K. (1958). "Ultraviolet absorption processes in the upper atmosphere." In *Advances in Geophysics*, Vol. 5, Academic, New York.

Watanabe, K., et al. (1967). "Absorption coefficient and photoionization yield of NO in the region 580–1350 Å." *Appl. Opt.* **6**, 391–396.

Watkins, W. R. and K. O. White (1977). "Water-vapor continuum absorption measurements (3.5–4.0 μm) using HDO-depleted water." *Opt. Lett.* **1**, 31–32.

Watkins, W. R., et al. (1979). "Water vapor absorption coefficients at HF laser wavelengths (2.64–2.93 μm)." *Appl. Opt.* **18**, 1582–1589.

Whiffen, D. H. (1966). *Spectroscopy*, Wiley, New York.

White, H. E. (1934). *Introduction to Atomic Spectra*, McGraw-Hill, New York.

White, K. O., et al. (1978). "Water vapor continuum absorption in the 3.5–4.0 μm region." *Appl. Opt.* **17**, 2711–2720.

Williams, C. S. and O. E. Becklund (1972). *Optics: A Short Course for Engineers and Scientists*, Wiley, New York.

Wolfe, W. L., ed. (1965). *Handbook of Military Infrared Technology,* Office of Naval Research, GPO, Washington, D.C.

Wolfe, W. L. and G. J. Zissis (1979). *The Infrared Handbook,* Environmental Research Institute of Michigan, Ann Arbor, Mich.

Worthing, A. G. (1939a). "Radiation laws describing the emission of photons by blackbodies." *J. Opt. Soc. Am.* **29,** 97–100.

Worthing, A. G. (1939b). "New λT relations for blackbody radiation." *J. Opt. Soc. Am.* **29,** 101–102.

Wyatt, C. L. (1977). *Theory and Methods of Radiometric Calibration,* Academic, New York.

Wyatt, P. J., et al. (1964). "The infrared transmittance of water vapor." *Appl. Opt.* **3,** 229–241.

Yale, G. D. et al. (1968). "The strengths of the N_2O bands near 4.5 μ." *Appl. Opt.* **7,** 695–697.

Yamada, H. Y. (1968). "Total radiances and equivalent widths of isolated lines with combined Doppler and collision broadened profiles." *J. Quant. Spectrosc. Radiat. Transfer* **8,** 1463–1473.

Yamamoto, G. (1961). "Numerical method for estimating the stratospheric temperature distribution from satellite measurements in the CO_2 band." *J. Meteorol.* **18,** 581.

Yates, H. W. and J. H. Taylor (1960). "Infrared transmission of the atmosphere." NRL Report 5453, June 8, 1960. NRL, Washington, D.C.

Young, C. (1965). "Tables for calculating the Voigt profile." College of Engineering Tech. Rpt. 05863-7-T, July 1965. (O.R.A. Project 05863) Dept. Aerospace Engineering, University of Michigan, Ann Arbor, Mich.

Young, C. and R.H. Bunner (1974). "Absorption of carbon dioxide 9.4 μm laser radiation by ozone. *Appl. Opt.* **13,** 1438–1443.

Zemansky, M. W. (1957). *Heat and Thermodynamics,* McGraw-Hill, New York.

Zissis, G. J. (1979). "Radiometry," in *The Infrared Handbook,* W. L. Wolfe and G. J. Zissis, eds., Environmental Research Institute of Michigan, Ann Arbor, Mich.

Zuev, V.E. (1974). *Propagation of Visible and Infrared, Radiation in the Atmosphere,* Halsted Press (Wiley), New York.

Author Index

Subject Index